Mathematical Models

Preliminary Edition

Robert Lee Kimball
Wake Technical Community College

PRENTICE HALL
Upper Saddle River, NJ 07458

Acquisitions Editor: Sally Denlow
Editorial Assistant: Joanne Wendelken
Editorial Director: Tim Bozik
Editor-in-Chief: Jerome Grant
Assistant Vice President of Production and Manufacturing: David W. Riccardi
Editorial/Production Supervision: Robert C. Walters
Managing Editor: Linda Mihatov Behrens
Executive Managing Editor: Kathleen Schiaparelli
Manufacturing Buyer: Alan Fischer
Manufacturing Manager: Trudy Pisciotti
Marketing Manager: Evan Girard
Creative Director: Paula Maylahn
Art Director: Jayne Conte
Cover Designer: Bruce Kenselaar

Printed in the United States of America

10 9 8 7 6 5 4 3 2

ISBN 0-13-271297-0

PRENTICE-HALL INTERNATIONAL (UK) LIMITED, LONDON
PRENTICE-HALL OF AUSTRALIA PTY. LIMITED, SYDNEY
PRENTICE-HALL CANADA INC. TORONTO
PRENTICE-HALL HISPANOAMERICANA, S.A., MEXICO
PRENTICE-HALL OF INDIA PRIVATE LIMITED, NEW DELHI
PRENTICE-HALL OF JAPAN, INC., TOKYO
SIMON & SCHUSTER ASIA PTE. LTD., SINGAPORE

Contents . . . . .Mathematical Models

1. . . . . **Numerical Investigations with Algebraic Models**
    1.1 . . . . Functions-A First Look
    1.2 . . . . Formulas
    1.3 . . . . .Investigating the Models

2. . . . . **Mathematical Modeling**

3. . . . . **Statistics-Dealing With Data**
    3.1 . . . . Sampling
    3.2 . . . . .Describing Data
    3.3 . . . . Experiments
    3.4 . . . . Groups Discussions and Work
    3.5 . . . . Line Graphs
    3.6 . . . . Using Technology

4 . . . .**Linear Functions**
    4.1 . . . . Relations and Functions
    4.2 . . . . .Linear Functions-The Slope
    4.3 . . . . Linear Functions-Intercepts and Applications
    4.4 . . . . Linear Functions-Finding the Mean
    4.5 . . . . More on Linear Functions-Applications and Models
    4.6 . . . . .Other Functions
    4.7 . . . . Electronic Spreadsheets-Taking the Pain Out of Functions

5 . . . .**More on Statistics-Linear Regression**
    5.1 . . . . .Measures of Central Tendency
    5.2 . . . . Standard Deviation
    5.3 . . . . Box and Whisker Plots
    5.4 . . . . .Linear Models of Best Fit
    5.5 . . . . Help! Using Spreadsheets for Lines of Best Fit

6 . . . . **Applications of Linear Equations**
    6.1 . . . . Systems of Linear Equations
    6.2 . . . . Solving Systems of Linear Equations
    6.3 . . . . Multiple Approaches to Systems of Linear Equations
    6.4 . . . . Matrices-An Introduction
    6.5 . . . . Matrices and Their Inverse
    6.6 . . . . Linear Programming - Inequalities
    6.7 . . . . Linear Programming - Introduction
    6.8 . . . . Linear Programming
    6.9 . . . . Linear Programming - More

7 . . . . **Applications of Probability**
    7.1 . . . . Introduction to Probability
    7.2 . . . . Area and Probability-Introducing the Normal Curve
    7.3 . . . . The Normal Curve-An Application of Probability
    7.4 . . . . .The Exponential Function

8 . . . . **Appendix**
    A Simulation
    Calculator Activities
    Ratios and Proportions
    A Numerical Study of Functions

A Note to the Student/Consumer

Mathematics has played a major role in world history. It has been used constructively in solving problems, explaining phenomena, and in developing ideas that improve our overall way of life. It has also been used to develop sophisticated weapons and improve our ability to defend our freedom.

The ability to use mathematics will be important to you. The citizen/worker needs to be able to use mathematics to make decisions and test conjectures. These skills do not replace the ability to perform basic arithmetic operations; they are in addition to them. Technology has changed what is expected of employees in the workplace as well as the amount of information available to the citizen/consumer. Employees no longer work in isolation; they work in teams in competition with others around the world. This global market and technologically advanced workplace has caused many changes in the business world. In reaction to the needs of the business world and of the citizen/consumer, the mathematics curriculum has been reformed. No longer are courses designed to upgrade algebraic skills the major requirement for an educated public. In fact, less advanced algebraic skills are needed in many of today's jobs because of technology. Technology allows you to perform calculations, investigate problems, and analyze data very easily. This text is a response to the changes occurring in the workplace and in the world of the consumer/citizen. The content you will study in this course will help you realize the *power of mathematics* and appreciate the *usefulness of mathematics as a tool*.

The mathematics needed for solving problems has often been developed by pure mathematicians who study a branch of mathematics building on what has already been proven, but with little thought to how it will ever be used. But sometimes, models are needed to explain physical phenomena or to predict results, so the mathematics was developed especially to solve these problems. In this text, you will investigate how models explain physical phenomena, use models to make predictions, and create models to solve problems. You will still need to learn some new mathematical skills, but they will be within the context of a larger problem.

In this text, you will explore several of the big ideas in mathematics. To be able to explore these ideas, you must have a level of mathematical sophistication that will enable you to use mathematics as a tool. The ideas presented in this text will help you solve real world problems through the development and exploration of mathematical models. The main task is to see how mathematics can be used to
- explain,
- predict, or
- make decisions.

If you have comments regarding what you learn in this course, I would welcome them! You may e-mail me at . . . . **rlkimbal@wtcc-gw.wake.tec.nc.us**

Robert L Kimball

How to Be Successful in This Class

A lot of your success will depend on both the quantity and the quality of your studying. Most students measure their studying in minutes or by the number of problems they accomplished. That is not always appropriate. Examine your productivity while studying mathematics. Did the problems you worked increase your knowledge and improve your skills at problem solving? Did the time you spent looking over notes and reading the text improve your understanding of a topic? Don't be fooled into thinking that if you can not work the first problem you can quit. That is when you really begin to work; not just harder, but smarter.

Here are some guidelines for success.
1. Attend class.

    When you begin work you will be expected to go to work everyday. Get in the practice of going to class everyday. Your instructor has been employed to explain the material for you. Often, the class is going to have experiences which can not be duplicated in texts. You will have to be present to be part of that learning experience.
2. Take good notes.

    Taking notes keeps you attentive and provides a guide to study by.
3. Read the text.

    Read the text slowly and carefully. Always read a mathematics text with a pencil, paper, and a calculator/computer. As you read and come to an example, cover up the solution and try to work it by yourself using the text if you get frustrated. Read key ideas several times to be sure you understand them before going on.
4. Do homework daily.

    Find study groups to work with. But, be careful. Study groups often give you a false impression of your abilities if you do not work the problem yourself. The people in study groups who benefit the most are the people who do the explaining. Always be sure you could explain a problem or concept to someone else.
5. Look over the material for the next day's lesson.

    Get a head start on learning the next day's work.
6. Do not be afraid to ask questions in class or to see your instructor.

    If you do not get a chance to ask a question in class or can not see your instructor, put the question in your portfolio. You might also try asking your question through e-mail. Get your instructor's address early in the quarter and it will prove beneficial for both of you.

On the first day of class <u>each week</u> turn in your *portfolio* to the instructor. At the end of the course, your portfolio will be a history of your progress. Problems you place in it should be neat and clearly identified (ie. Section 3.2 - Problem 4). You should obtain some sort of folder in which this work is to be placed. Keep all the work in the folder as you add new work. When the portfolio is turned in, identify the work to be graded by using a paper clip. The portfolio should contain the selected homework for the previous week and your journal entry for the previous week (see below). The portfolio will be returned to you on the next class meeting.

Portfolio Guide

<u>Journal Entry</u>

At the end of <u>each </u>week during the quarter write two paragraphs about that week's classes. If for some reason the portfolio is not turned in, there should still be two paragraphs for that week's work.

In the *first paragraph* include a description of the topics that you covered and the type of problems that were assigned that week. Also include the amount of time (in hours?) you spent doing homework and studying for this class; and, most importantly, include your feelings about the class.

In the *second paragraph*, answer the following questions:
> What did you like best about the class this week?
> What did you like least about the class this week?
> How is school going for you?
> If you could have changed one thing about this week, what would it have been?

Your efforts toward developing a portfolio will be very rewarding. Your instructor is concerned about your progress in the course and about your progress toward your educational goals. Your journal entries will allow you both to get to know one another better.

<u>Selected Homework Assignments</u>

☺ Include the problems marked with a "smiley face" in your Portfolio.

To the Teacher/Coach

I hope *Mathematical Models* will provide you with some of the resources needed to provide your students with a mathematics course they will find interesting and one that will prove useful. If students are not motivated to study mathematics further, I hope they will have experienced some of the power mathematics provides the problem solver. The content should provide the foundation for further study in the mathematics classroom or in the workplace. The AMATYC document *Crossroads in Mathematics, Standards for Introductory College Mathematics Before Calculus* calls for all students to experience the power of mathematics through a college-level mathematical experience. It also calls for classrooms to be student-centered and for learning to be an active experience. The material in this text should provide the resources to allow you to follow those guidelines.

The text emphasizes the need to be able to examine a verbal problem symbolically, numerically or graphically. It often refers to the spreadsheet as an appropriate tool for these investigations. Students taking this class who own graphing calculators should be encouraged to use them. Graphing calculators also have the ability to investigate functions numerically and graphically but they are not as prevalent in the workplace. Since the spreadsheet will likely be used in other classes the student will take and in the workplace, the spreadsheet was chosen as the technology of choice for this course. The applications in the text should be meaningful and easily extended for further study. The sections are written in the context of an application and often include other activities which can be worked on collaboratively. A goal of this text is not to provide enough examples so that the student can work ten more like them, but rather provide examples which promote discussion and thought and which provide a positive learning experience which can be transferred to other problem solving opportunities. Chapter One provides a clear indication that this course is going to be different. There are a number of ways you might coach the students through this chapter. I think there are several parts that all students should experience (A, B, E, J, & K). Other parts could be assigned to groups or individually. This chapter provides students opportunities to review basic skills and an initial look at several important ideas. All in all, most students should find this mathematics course different from any they have taken before. Finally, I hope you will use the portfolio outline suggested. I have found that students willingly share information with you to make this class a better experience for you and for them.

Rob Kimball

Chapter 1   Numerical Investigations with Algebraic Models

This chapter will demonstrate the usefulness of mathematics. It will also examine some of the tools you, as a mathematician, will use to find solutions to problems. In order to use mathematics as a tool in solving practical problems, a mathematical model must be developed that describes the real-world situation. The formulas used in this chapter are examples of mathematical models that can be used to explain, predict and make decisions regarding the situation.

The formulas discussed in this chapter will allow you to practice evaluating expressions using fractions, decimals, exponents and the appropriate order of operations. Spreadsheets may also be utilized to examine these phenomena.

1.1   Functions

Objectives of this section. The student will be able to
- *define: function, formula, independent variable and dependent variable*
- *identify three ways of defining functions*
- *recognize and understand functional notation.*

Businesses are interested now, more than ever, in conserving the Earth's natural resources and saving money. The recycling industry has grown to meet this demand. One large company produces a product that generates a great deal of unused sheet metal strips about 1-inch wide and 8-inches long. These strips are placed in large 55-gallon drums and delivered to a recycling business. The drums are hardly ever filled to capacity since each drum is the result of one day's work. The company uses the following formula to estimate the weight of each drum and its contents before it is shipped.

$$y \ = \ 22 + 40x$$

What do you suppose the x stands for?

Formulas

The total weight of the 55-gallon drum is certainly *related* to the height to which the drum is filled. You might say that the total weight "is a function of" the height to which the drum is filled. The dependence which exists in this example and in others like this is a very important idea in mathematics. In mathematics we say that the total weight is a **function** of the height to which the drum is filled. A **formula** explicitly informs the user of that relationship. The formula for the total weight

$$y \ = \ 22 + 40x$$

demonstrates that *the total weight, **y**, (measured in pounds) is equal to 22 pounds plus 40 pounds per inch of height, **x**.* The formula adds pounds to pounds to get pounds. An analysis of the units is very important when using formulas.

There are many formulas found in geometry. The following figures provide the formulas for the perimeter and area of plane figures and the surface area and volume of three-dimensional figures.

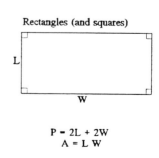

Rectangles (and squares)

P = 2L + 2W
A = L W

The values of x and y change with different barrels. For this reason *x and y* are called *variables*. The *x* is the **independent variable** and the *y* is the **dependent variable.** The formula takes an **input, x** and provides an **output, y** according to the relationship expressed in the formula. Although the values of the variables change, the relationship remains unchanged.

If the product being placed in the barrels changed, then the relationship between the variables x and y would be different. If the material was more dense, the formula might be

$$y = 22 + 56x$$

If the material was less dense, the formula might be

Formulas may express a relationship between two variables; however, some express relationships between more than two variables (see the examples to the right). The letter used in formulas as well as functions is completely arbitrary. The examples in this text will use letters which make sense in the problem.

To find a y-value for any x-value, simply substitute the x-value into the formula for x and compute. Complete the table using the formula for the total weight of sheet metal strips shown above.

x-value	0	10	20	30	40
y-value					

Table for Total Weight of a 55-gallon Drum

The **table** you have just completed uses arbitrarily chosen x-values to obtain y-values. Together, the x- and y-values produce **ordered pairs** of values.

Tables

In the example above, the table was constructed after the formula describing the relationship between x and y was known. This is not always the case. Sometimes, either because the relationship isn't known or because there is not a mathematical relationship, the table must come first. The table itself defines the function and is the only information available. Many real-world functions are determined by collecting data and arranging the data into tables.

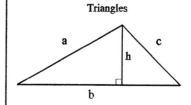
Triangles

$$P = a + b + c$$
$$A = \tfrac{1}{2}bh$$

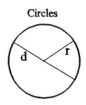
Circles

$$C = 2\pi r = \pi d$$
$$A = \pi r^2$$

Rectangular Solid

$$S = 2ab + 2bc + 2ac$$
$$V = abc$$

Right Circular Cone

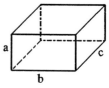

$$S = 2\pi r^2 + 2\pi rh$$
$$V = \tfrac{1}{3}Ah = \tfrac{1}{3}\pi r^2 h$$

Sphere

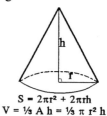

$$S = 4\pi r^2$$
$$V = 4/3\,\pi r^3$$

The table shown here defines a relationship between the date and the total sales for a retail establishment. Notice there is still an input variable (the date) and an output variable (the total sales).

A mathematician would examine the data in the table and determine if a formula exists to **model** the relationship by looking for patterns in the data. If a formula exists, then it can be used to find an output value for any input value whether or not the value is in the table.

Consider some information that you see regularly: the daily high temperature. The high temperature for the last several days is available information. Would a model obtained from that data be able to *predict* future high temperatures? The Dow Jones Average is shown daily in the newspaper. Would a model obtained from that data be able to *predict* the average next July?

This course includes information on techniques for analyzing functions defined by data. But many functions cannot easily be described in the convenient shorthand of a mathematical formula. They must be defined using a table or a graph.

Graphs

In addition to algebraic formulas and tables of values, a graph may also be used to depict a relationship between two variables. Often graphs are used to display data from a table. Graphs allow us to see **trends** in the data that might not be easily seen in a table.

Graphs are constructed by placing the *ordered pairs* from tables onto a **coordinate axis.** The coordinate axis is made by drawing two lines at right angles. The point of intersection, the **origin**, has coordinates **(0, 0)**. Beginning with the origin, the four segments are scaled appropriately.

Graphs will be shown in this text in which individual data points are graphed (*scatter plots*) and those in which the data points are connected by a line (*line graph*).

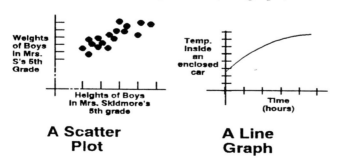

A Scatter Plot   **A Line Graph**

Date	Total Sales
Oct 4 Friday	$2550
Oct 5 Saturday	$3350
Oct 6 Sunday	$3210
Oct 7 Monday	$1020
Oct 8 Tuesday	$808
Oct 9 Wednesday	$1300
Oct 10 Thursday	$1250
Oct 11 Friday	$2300
Oct 12 Saturday	$3880
Oct 13 Sunday	$3300
Oct 14 Monday	$1100
Oct 15 Tuesday	$1000

Functional Notation

Functions play such a central role in mathematics that they have their own notation. If a relationship exists between two variables, say x and y, then we demonstrate that *y is a function of x* by using the following notation:

$$f(x) = 22 + 40x$$

This is read, "*f of x is 22 plus 40 times x*".

The notation *f(x)* indicates that the value of the function *f* depends on the value of the independent variable *x*.

f(20) indicates the value of the function *f* when x is 20.

$$f(x) = 22 + 40x$$
$$f(20) = 22 + 40(20)$$
$$f(20) = 22 + 800$$
$$f(20) = 822$$

☺1.   Read the section and write a definition for every term that is **bold-face** type. You should also have a good understanding of the terms in *italics*. This is good practice for <u>every</u> section in the text!

2.   Consider the formula for the area of a circle.
     a.   How many variables are in this formula?
     b.   Identify (describe in words) the independent variable and the dependent variable.
     c.   Express this relationship in words.
     d.   If the independent variable has a value of 3 inches, what value does the function assign to A?
     e.   Write the formula in functional notation. Find A(5.5 inches).
     f.   Is there a reasonable output for an input of -3?

3.   The following graph represents a functional relationship between the amount of time since death occurred and the temperature of a corpse. The x-axis is scaled in units of 1 hour, and the y-axis is scaled in units of 10 degrees fahrenheit.

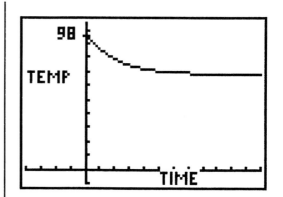

     a.   About how many hours does it take for the corpse to reach room temperature?

     b.   At about what temperature does the corpse seem to reach and remain?

4.   The following table shows the population of the Triangle (an area between and around Raleigh, Durham and Chapel Hill, NC) in 1976 and 1996. This area has consistently been among the fastest growth areas in the country and was ranked #1 in places to live in 1995. During the past few years the population has grown at a faster pace than ever before.

Year	Population of the Triangle
1976	616,700
1996	1,014,374

     a.   Which of the three models shown in the graphs to the right do you think is the best to model the population growth in the Triangle area?

     b.   Discuss why you think you can or cannot predict the population of this area in the year 2004?

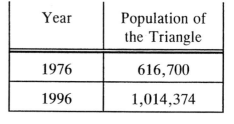

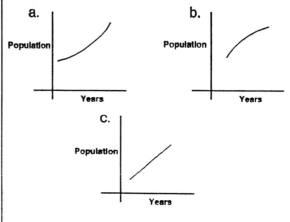

1.2   Formulas

Objectives of this section. The student will be able to
- *appreciate the use of formulas as mathematical models*
- *evaluate a formula*
- *understand what the result of a calculation means*
- *review basic arithmetic and algebraic skills in the context of a real application*
- *build teamwork skills*

Mathematical calculations are a part of life. Formulas or mathematical models can be used to find unknown values. These formulas demonstrate a relationship between two or more unknowns or variables. In order to use the formulas represented here, you may have to review the basic concepts of applied arithmetic and/or algebra.

Your instructor may teach this section in several ways. Groups may work through an application and report their work to the class. Some applications may be assigned as individual work and some may be discussed by the instructor. The student is encouraged to examine as many applications as possible. You should "work through" not read the applications. Take the time to strengthen the basic skills used in each application.

A.   The Tax Rate Table

The 1985 North Carolina Tax Forms (D-400) included a formula for finding the amount of tax owed based on Net Taxable Income. Therefore, this formula relates the two variables, Net Taxable Income (NTI), and Tax Owed (T). This chart is from NC D-400 (1985). It allows you to compute the Tax Owed based on the amount of Net Taxable Income.

Find the amount of Tax Owed if the NTI = $7805.

TAX RATES

If Net Taxable Income (line 15) is

over	but not over	The Tax is
-0-	$2000	3% of net taxable income
$2000	$4000	$60+4% of amt. over $2000
$4000	$6000	$140+5% of amt. over $4000
$6000	$10000	$240+6% of amt. over $6000
$10000		$480+7% of amt. over $10000

NC Tax Rate Table (1985)

Applications of percents:

What is 5% of $14,400?

This question translates to an algebraic equation:
$$x = .05 (\$14400)$$
To solve for x simply multiply:
$$x = \$720$$

5% of what number is $255?

This question also translates to an algebraic equation:
$$.05 (x) = \$255$$
To solve for x simply divide both sides by .05:
$$\frac{.05 (x)}{.05} = \frac{\$255}{.05}$$
$$x = \$5100$$

a) What is 7% of $22000?
b) 9% of what number is $555?

Since $7805 is over $6000 ($7805 > $6000) but not over $10,000 ($7805 ≤ $10000), then the Tax Owed is $240 + 6% of the amount over $6000. This statement must be translated into a mathematical formula. The key words are *is* which means equal to, *of* which means multiplication and *amount over* which means difference or subtraction. The result is the following formula:

T = $240 + .06 (NTI - $6000)

Solving the problem requires you to substitute for the known value, NTI, and find the unknown value, T:

$$T = \$240 + .06 (\$7805 - \$6000)$$
$$T = \$240 + .06 (\$1805)$$
$$T = \$240 + \$108.30$$
$$T = \$348.30$$

It is customary to round to the nearest dollar on tax forms.

> *To change a percent to a decimal you move the decimal over two places to the **left**.*

> It is customary to use single letters like x or y as the variables in algebra, but that is not necessary. In these models we will use variables which make sense to the problem--sometimes using more than one letter.

> *According to the order of operations, you do what is inside parenthesis, perform any operations with exponents, then do multiplication and division and then do addition and subtraction.*

Answer: *The amount of tax owed is $348.00*

```
240+.06(7805-600
0)
        348.3000000
```

Extending the Problem

You may be asking why there were five different categories of taxable income in 1985. That is, why is there a category for people who made over $10,000, another for those who made between $6,000 and $10,000, and so on. The answer can clearly be seen in a *graph* of this model.

Graphs

Graphs commonly aid in the analysis of a model. This model relates two variables and thus we need to use a two-dimensional picture for this model. The coordinate axis shown here is used to graph relationships between two variables. It is customary to call these two variables x *(the independent variable) and* y *(the dependent variable)*. In this model our independent variable is **NTI** and the dependent variable is **T**.

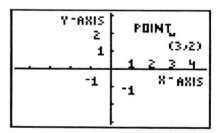

As a review, graph the following points using the figure shown here. Connect the points by drawing a line through them.

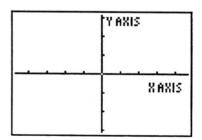

POINTS: (-3, -1), (-1, 0), (1, 1), (3, 2)

To construct a graph of the 1985 Tax Table model, first create ordered pairs in the form of *(NTI, T)* and connect the ordered pairs with a line. These skills will be developed later in the course. For now, the graph is simply shown here.

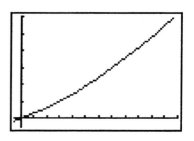

Figure A

It is not easy to see the affect that changing the percentage in each category has on the graph in Figure A. Figure A shows the Tax Owed for Net Taxable Income from $0 to $12500 in increments of $1000. But, if you look closely, you see a slightly upward slant in the "line" as you move from left to right (as the NTI increases).

Figure B looks more closely at only the first three categories by changing the horizontal scale to show only values between $0 and $6000 in increments of $1000. The two vertical lines (at $2000 and at $4000)

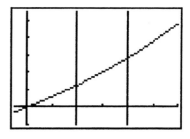

Figure B

separate the graph into three regions analogous to the first three tax categories in the Tax Rate Table. In this figure you can see clearly that the slope of the *tax-line* for NTI between $4000 and $6000 is steeper than for NTI between $2000 and $4000 which is steeper than for NTI between $0 and $2000.

Numerical Investigations

Models may also be examined using numerical methods. The following tables demonstrate the effect that the 3% rate of taxation has on incomes between $0 and $2,000, the effect the 4% rate of taxation has on incomes between $2,000 and $4,000 and the effect the 5% rate of taxation has on incomes between $4,000 and $6,000. These *numbers* further clarify the *graph* and interpret the *symbolic model*.

Category 1

NTI	0	400.00	800.00	1,200.00	1,600.00	2,000.00
Tax Owed	0.00	12.00	24.00	36.00	48.00	60.00
Increase in Tax Owed		12.00	12.00	12.00	12.00	12.00

Category 2

NTI	2000.00	2,400.00	2,800.00	3,200.00	3,600.00	4,000.00
Tax Owed	140.00	156.00	172.00	188.00	204.00	220.00
Increase in Tax Owed		16.00	16.00	16.00	16.00	16.00

Category 3

NTI	4000	4,400.00	4,800.00	5,200.00	5,600.00	6,000.00
Tax Owed	340.00	360.00	380.00	400.00	420.00	440.00
Increase in Tax Owed		20.00	20.00	20.00	20.00	20.00

The numbers in the last row of each table demonstrate the affect the percentage has on the tax owed. Each $400 increase in NTI causes a $12 increase in taxes in category 1, a $16 increase in category 2 and a $20 increase in category 3.

Explain in *words* what this means to the taxpayer!

B.   <u>Base Salary Plus Commission</u>

When Jake started working with a company that sold copiers, he wanted some assurance that his income would be enough to cover his living expenses. He also wanted to be able to earn a salary based on his sales performance. The company and Jake mutually agreed that Jake's salary would be $1000 per month plus an 8½% commission of all sales over $15,000 in that month. If his sales don't exceed $15,000 for any one month, he still gets $1000. This arrangement relates the Total Sales (TS) to Jake's monthly Salary (S). His monthly Salary is dependent on his Total Sales. This arrangement translates to the following formula:

$$S = \$1000 + .085 \ (\ TS - \$15,000)$$
$$[TS > \$15,000]$$

$$S = \$1000 \quad [TS \leq \$15,000]$$

To change 8½% to a decimal, you first change the ½ to a decimal: .5. That means that 8½% is equal to 8.5%. Now move the decimal two places to the left:

.08.5%

The symbol > means "greater than" or "more than." If x > 5 then x represents all the numbers greater than 5:

The symbol < means "less than." If z < 3 then z represents all the numbers less than 3:

The symbol ≥ means "greater than or equal to" and the symbol ≤ means "less than or equal to."

Find Jake's monthly Salary if his Total Sales for that month were $22,505.

Since his sales exceed $15,000 ($22,505 > $15,000) we will use the first formula.

$$Salary = \$1000 + .085 \ (\ \$22,205 - \$15,000 \ )$$

$$Salary = \$1000 + .085 \ (\ \$7,205 \ )$$

$$Salary = \$1000 + \$612.42$$

$$Salary = \$1612.42$$

Answer: *Jake's salary for this month is $1612.42.*

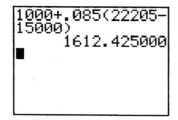

C.   The Winner's Share

East High School is having a fund raiser. The goal is to raise $1000 for new computer software for the mathematics faculty. Tickets are to be sold for $2 each. The school expects to sell at least 1000 tickets. Each ticket will allow the buyer to write three numbers on the ticket and the same three numbers on the stub. If these three numbers match the three numbers picked at random, the holder of that ticket wins! (See the instructions on the ticket.)

Ticket Stub	East High School
5 7 3	**Software Fund Raiser $2**
___ ___ ___	**5 7 3**
Name: *Adam Lofton*	___ ___ ___
Phone Number: 876-5432	Instructions: Place three integers in the three spaces on this ticket and on the attached stub. If your 3 numbers exactly match the 3 numbers drawn at random by the math faculty - you win. The amount you win is based on the number of people who enter and on the number of winners. Details are available.

The amount of revenue collected on $2 tickets depends on the number sold. If 5 were sold the revenue would be
$$\$2 \times 5 = \$10$$
If 6 were sold:
$$\$2 \times 6 = \$12$$
If 7 were sold:
$$\$2 \times 7 = \$14$$
So, if the variable *TS* represents the number of tickets sold, then the revenue generated is
$$\$2 \times TS$$
The "profit" is the revenue minus the cost (or expenses):
$$\$2 \times TS - \$1000$$
Write an expression for the profit is *n* tickets are sold at $4 each and the cost is $1200.

Here are the details. There may be more than one winner. It is obviously possible that more than one person will write down the same three numbers as the math faculty select, so more than one person will have a "winning" ticket. The plan is to keep $1000 for software and divide up what is left among the winners. Therefore, the Individual Prize Money (IPM) depends on both the Number of Winners (W) and the Number of Tickets Sold (TS).

In division the line that separates the numerator from the denominator is a grouping symbol. That line means that you do everything on top and everything on the bottom, and then divide.

The formula is

$$IPM = \frac{2(TS) - \$1000}{W}$$

Find the Individual Prize Money if the school sells 1741 tickets and three winners come forward.

$$IPM = \frac{2(1741) - \$1000}{3}$$

$$IPM = \frac{\$3482 - \$1000}{3}$$

$$IPM = \frac{\$2482}{3} = \$827.33$$

```
(2*1741-1000)/3
         827.3333333
```

Answer: *Each of the three winners would get $827.33 leaving the school $1000.01 to spend on software.*

D.   Future Value

A banker needs to calculate the Future Value (F) of an Investment (P) given the Periodic Interest Rate (r) and the Number of Periods (t). The formula is known to be

$$F = P(1 + r)^t$$

In this formula the Future Value (F) depends on three other variables: P, r and t.   Find the Future Value of a $1000 investment that earns interest at 6% compounded monthly over 5 years. The interest period is monthly which means that there are 12 periods in each year.   Therefore the monthly rate is .06/12 = .005 and there are 60 periods in five years.

> An exponent means that you multiply the base that many times itself.
>
> Exponent, Base: $5^3 = 5 \times 5 \times 5 = 125$
>
> When the exponent is too large to calculate using regular multiplication, then a calculator is used. To find
>
> $$1.006^{48}$$
>
> | 1.006 | y^x | 48 | $=$ | 1.33261 |

$$F = \$1000(1 + .005)^{60}$$

$$F = \$1000(1.005)^{60}$$

$$F = \$1000(1.348850153)$$

$$F = \$1348.85$$

Answer: *The future value of $1000 invested at 6% compounded monthly over five years is $1348.85.*

Compute with the aid of a calculator or spreadsheet:
a) 3^4
b) 1.3^4
c) 1.03^4
d) 1.03^{40}

```
1000(1+.06/12)^6
0
      1348.850153
```

Extending the Problem

The following is an example of a *non-linear* model. Obviously, non-linear means that the model is not linear. A graph will be employed to demonstrate what non-linear means. This graph shows how the *future value* depends on the amount of *time* the investment is allowed to grow if the *principal* and *interest rate* are fixed.

Figure C shows the value of an investment monthly given an initial investment of $1000 at an interest rate of .05% monthly.   Figure D shows the value of an investment monthly given an initial investment of $1000 at an interest rate of .06% monthly.

Explain why you would like your interest to accrue at .05% per month rather than at .06%. (Use the graph in your explanation.)

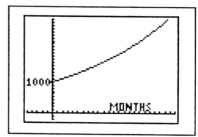

Figure C: at .05%

Figure D: at .06%

E.   The New Room

A family has built a new addition to its home, a 14 ft. 6 in. × 18 ft. 3 in. family room.   They need to make two calculations: the area of the room and the perimeter of the room. They need the area to get a price on hardwood floor and the perimeter for ceiling molding.   The hardwood floors are $6.50 per square foot (rounded up to the nearest integer) and the ceiling molding is $4.25 per linear foot (sold only in 8 foot lengths). How much would these two items cost?

When making calculations with units it is important that the operations be performed on numbers with the same units. That is, add feet to feet or multiply inches times inches. In this problem the inches were changed to feet by making them a fraction of a foot. Six inches is six twelfths of a foot, 3 inches is 3 twelfths of a foot.

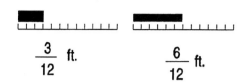

$\dfrac{3}{12}$ ft. $\dfrac{6}{12}$ ft.

The formula for the Perimeter (P) of a rectangle (or any polygon) is the distance around the figure. Since the two opposite sides in a rectangle are the same, the perimeter is based on two dimensions: the length (L) and the width (W). The formula for the perimeter of a rectangle is

$$P = 2(L) + 2(W)$$

The formula for the area (A) of a rectangle is the length times the width:

$$A = (L)(W)$$

The area and perimeter calculations are shown here.

$$P = 2(14 + \tfrac{6}{12} \text{ ft.}) + 2(18 + \tfrac{3}{12} \text{ ft.})$$

$$P = 29 \text{ ft.} + 36.5 \text{ ft.}$$

$$P = 65.5 ft$$

$$A = (18 + \tfrac{6}{12} \text{ ft.})(14 + \tfrac{3}{12} \text{ ft.})$$

$$A = (18.5 \text{ ft.})(14.25 \text{ ft.})$$

$$A = 263.625 \text{ square feet}$$

A mixed number is changed to an improper fraction by multiplying the whole number by the denominator and adding the numerator.

4⅜ = 35/8

Convert to an improper fraction:
a) 3 ⅞
b) 2 ¾

Multiplication of fractions is performed by multiplying the numerators and then multiplying the denominators.

$$\frac{5}{12} \times \frac{7}{12} =$$

$$\frac{5 \cdot 7}{12 \cdot 12} = \frac{35}{144}$$

Find: a) $\dfrac{5}{16} \times \dfrac{3}{16}$

b) $\dfrac{1}{2} \times \dfrac{3}{8}$

The total cost is

$$Cost = 72 \text{ ft.} \times \$4.25 \text{ per ft.}$$
$$+ 264 \text{ sq.ft.} \times \$6.50 \text{ per sq.ft.}$$

$$Cost = \$306 + \$1716.00 = \$2022.00$$

Answer: *The two projects would cost $2022.00.*

```
2(14+6/12)+2(18+
3/12)
        65.50000000
(18+6/12)*(14+3/
12)
        263.6250000
```

F.   The Bullet's Path

A bullet is a projectile that travels on a relatively straight path. Therefore, the point of origin of a bullet may be estimated if two points along the path of the projectile are known. The shorter the distance and more linear the path, the greater the accuracy of the prediction. Similar triangles and the proportions that result from two triangles being similar provide the mathematics to perform these calculations.

The figure below shows the path of a bullet through points **A** and **B** originating at some point **C** located in another building. Using the point that the bullet entered the building (**B**) and the point that it left the building (**A**), a hypothesis can be made concerning the window from which the bullet was fired in the adjacent building.

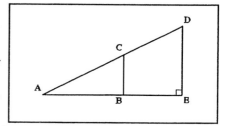

$\triangle ABC \sim \triangle AED$

That means that

$$\frac{AB}{AE} = \frac{AC}{AD}$$

and

$$\frac{BC}{ED} = \frac{AC}{AD}$$

and

$$\frac{BC}{AB} = \frac{ED}{AE}$$

These statements are called proportions. Proportions may be solved by multiplying the numbers on the diagonals together resulting in a simple equation. The three examples illustrate this concept.

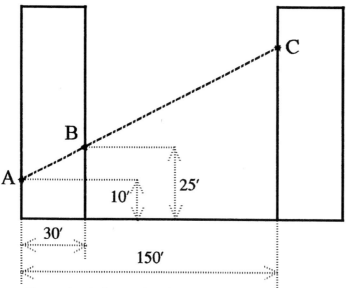

From the information in the figure, a proportion can be written and solved.

$$\frac{15 \text{ ft.}}{30 \text{ ft.}} = \frac{x}{150 \text{ ft.}}$$

$$\rightarrow (30 \text{ ft.})(x) = (15 \text{ ft.})(150 \text{ ft.})$$

$$\rightarrow x = 75 \text{ ft.}$$

Answer: *The bullet came from a window approximately 85 feet above the street level.*

If

$$\frac{2}{3} = \frac{x}{12}$$

$$\rightarrow 3(x) = 2(12)$$

$$\rightarrow 3x = 24$$

$$\rightarrow x = 8$$

If

$$\frac{5}{x} = \frac{7}{11}$$

$$\rightarrow 7(x) = 5(11)$$

$$\rightarrow 7x = 55$$

$$\rightarrow x = 7.85714$$

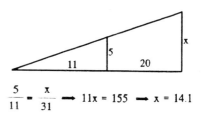

$$\frac{5}{11} = \frac{x}{31} \rightarrow 11x = 155 \rightarrow x = 14.1$$

Extending the Problem

Triangles appear in many applications. Two other facts about triangles are useful and will be presented here.

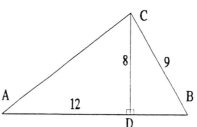

Right Triangles

The *Pythagorean Theorem* states that there is a relationship between the three sides of any right triangle as stated in the figure to the right. The basic identity $A^2 + B^2 = C^2$ can then be solved for each variable producing three other identities. Using this theorem, the missing side may be found if any two of the other sides are given. The three examples below illustrate this fact.

$$A^2 + B^2 = C^2$$

$$A = \sqrt{C^2 - B^2}$$

$$B = \sqrt{C^2 - A^2}$$

$$C = \sqrt{A^2 + B^2}$$

$A = 3.0$
$B = 8.0$
$C = ?$

$A = 5.3$
$C = 8.6$
$B = ?$

$C = 6.0$
$B = 2.4$
$A = ?$

$C = \sqrt{3^2 + 8^2}$
$ = \sqrt{73}$
$ = 8.5$

$B = \sqrt{8.6^2 - 5.3^2}$
$ = \sqrt{45.9}$
$ = 6.8$

$C = \sqrt{6.0^2 - 2.4^2}$
$ = \sqrt{30.2}$
$ = 5.5$

All Triangles

The Pythagorean Theorem is used to calculate the length of a missing side in a <u>right</u> triangle. If the area of a triangle is needed then another theorem is useful. Hero's formula provides a means of finding the area of a triangle given the length of all three sides. The example below shows how to use Hero's Formula to calculate the area of a triangle.

Hero's Formula:

$$A = \sqrt{s\,(s - x)(s - y)(s - z)}$$

where x, y, and z are the lengths of the three sides of any triangle, and

$$s = \frac{x + y + z}{2}$$

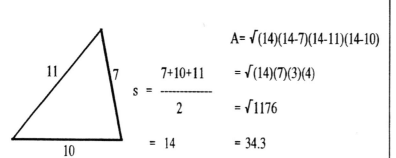

$s = \dfrac{7+10+11}{2}$

$ = 14$

$A = \sqrt{(14)(14-7)(14-11)(14-10)}$

$ = \sqrt{(14)(7)(3)(4)}$

$ = \sqrt{1176}$

$ = 34.3$

Find the area of △ABC two ways:
○ using the sum of the two small triangles and
○ using Hero's formula

G. **Straight Line Depreciation**

A business is allowed to depreciate its long-term assets over several years. Several methods of depreciation are allowed. The straight-line method allows the owner to depreciate an equal amount over the "life" of the asset. The final value of the asset is known as the "scrap value." The annual depreciation (AD) is found using the following formula based on the initial value (IV) and the scrap value (SV) and the number *(n)* of years representing the life of the asset.

$$AD = \frac{IV - SV}{n}$$

Find the Annual Depreciation if the initial value was $13,050, the scrap value is $1000 and if the life of the asset is 7 years.

$$AD = \frac{\$13050 - \$1000}{7}$$

$$AD = \$\frac{12050}{7}$$

$$AD = \$1721.43$$

Answer: *The annual depreciation is $1721.43.*

If the annual depreciation remains constant each year, then there is a constant change in one variable with respect to another. That is, the ratio of the change in **y** over the change in **x** is constant. For example, a depreciation of $200 per year is a ratio:

$$\frac{\$200}{1 \text{ year}}$$

This ratio represents the change in y over the change in x.

To calculate the depreciation for five years, solve the proportion by multiplying along the diagonals:

$$\frac{\$200}{1 \text{ year}} = \frac{Depreciation}{5 \text{ years}}$$

$$Depreciation = 5 \text{ years} \cdot \frac{\$200}{1 \text{ year}}$$

$$Depreciation = \$1000$$

Find x:

a) $\frac{x}{5} = \frac{3}{4}$

b) $\frac{5}{12} = \frac{x}{4}$

c) $\frac{3}{8} = \frac{9}{x}$

```
(13050-1000)/7
         1721.428571
■
```

H. **Monthly Payments**

Betty is going to have to borrow some money in order to buy a car she needs. She has made arrangements to borrow $8600 at an APR (Annual Percentage Rate) of 7.98% for 48 months. What will her monthly payment be?

Recall, to change a percent to a decimal, move the decimal two places to the left.

7.98% = .0798

The amount of the Monthly Payment (M) necessary to repay a loan with interest is dependent on the Amount Borrowed (P), the Monthly Interest Rate (r) and the number of Months (t) according to the formula:

$$M = \frac{Pr(1 + r)^t}{(1 + r)^t - 1}$$

The order of operations is crucial here. The monthly interest rate must be added to one (perform the operations inside the parenthesis first) and then the exponentiation is performed. The multiplication is performed in the numerator and then subtraction performed in the denominator. Then, the division is performed.

The periodic interest rate is found by dividing the APR by 12. Thus $r = .0798/12 = .00665$. Substituting these values into the formula

$$M = \frac{(\$7800)(.00665)(1 + .00665)^{48}}{(1 + .00665)^{48} - 1}$$

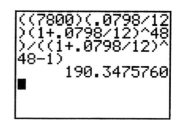

$$M = \frac{(\$7800)(.00665)(1.00665)^{48}}{(1.00665)^{48} - 1}$$

$$M = \frac{(\$7800)(.00665)(1.3745733)}{.3745733}$$

$$M = \frac{\$71.2991161}{.3745733}$$

Answer: *The monthly payment needed to amortize this loan is $190.35.*

$$M = \$190.35$$

I.   <u>True Cost</u>
When Lee got his license, he decided he would rather pay for using his Dad's car based on the number of miles driven rather than buying his own car. So he and his Dad sat down and listed all the costs associated with driving a car in order to find a model that could be used to find the cost per mile when Lee used his Dad's car. If his Dad's car is driven 24,000 miles annually, find the cost per mile for Lee to use his Dad's car.

This list shows the items with the associated cost.

Depreciation	$2500 per year
Tires	$250 every 40,000 miles
Gasoline	$.06 every mile
Oil	$20 every 3000 miles
Automobile Insurance	$1100 per year
Taxes	$120 per year
Expenses	$500 per year

They discovered that some costs were annual costs and some were dependent on the number of miles driven. The Cost per mile (C) is dependent on the number of miles driven each Year (Y) and the number of miles driven by Lee (L).

$$C = \frac{L(\$2500 + \$1100 + \$120 + \$500)}{Y} + \$.06L + \frac{\$20L}{3000} + \frac{\$250L}{40000}$$

$$C = \frac{\$4220L}{Y} + \$.07292L$$

Substituting $Y = 24000$ miles and $L = 1$ the cost per mile for Lee to use his Dad's car is found.

$$C = \frac{\$4220}{24000} + \$.07292$$

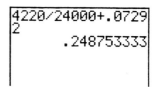

$$C = \$.24875 = \$.25$$

Answer: *Lee would have to pay 25 cents per mile to use his Dad's car.*

J.   Distance, Rate and Time

As an object moves, the distance, rate and time it travels are related by the relationship shown here.

D = R T

When any two of the three values are known, the third can be found.

When two objects are moving along the same path and in the same direction, then the relative speed of the two objects is the *difference in their speeds.* That is, if two cars are moving down a four-lane road, one at 60 mph and the other at 50 mph, then the first car is traveling 10 mph faster than the other.

When two objects are moving along the same path and in opposite directions, then the relative speed of the two objects is the *sum of their speeds.* That is, if two cars are traveling toward each other, one at 60 mph and the other at 50 mph, then the speed at which they are traveling toward one another is 110 mph.

Ex. 1 Two cars traveling down a four-lane interstate begin side by side. One averages 55 mph and the other averages 70 mph. How far apart will they be after 100 minutes?

Ex. 2 Two cars are headed toward one another on a two-lane road. One is traveling at 45 mph and the other at 52 mph. After they pass one another, how long will it be (in seconds) before the two cars are 2000 feet apart?

Example 1

$$D = R T$$
$$= (70 - 55 \tfrac{miles}{hour})(\tfrac{100}{60} \ hours)$$
$$= (15 \tfrac{miles}{hour})(\tfrac{5}{3} \ hour)$$
$$= 25 \ miles$$

Example 2

$$T = \frac{D}{R}$$
$$= \frac{2000 \ ft}{(45 + 52 \tfrac{miles}{hour})}$$
$$= \frac{2000 \ ft.}{(97 \tfrac{miles}{hour})}$$
$$= \frac{2000 \ ft.}{142.3 \tfrac{feet}{second}}$$
$$= 14.1 \ seconds$$

Recall

Distance, Rate and Time

If   D = R T,

then   R = D/T

and   T = D/R

1) Find the distance traveled if an object travels at 30 ft/sec for 5 seconds.

2) Find the average rate at which an object traveled if it traveled 450 miles in 15 hours.

3) Find the time it took to for an object to travel 300 meters at a rate of 50 km/hr.

Converting Units

$$\text{To change } \tfrac{x \ ft}{sec} \text{ to } \tfrac{y \ miles}{hour}:$$

$$\frac{x \ ft.}{sec} \times \frac{1 \ mile}{5280 \ ft} \times \frac{3600 \ sec}{1 \ hour} = \frac{y \ miles}{hour}$$

In example 1 it is necessary to convert minutes to hours in order for the units to cancel.

In example 2 it is necessary to convert miles per hour to feet per second since the distance is given in feet. (The 2000 feet may just as easily been converted to miles but the time would then have been in fractions of an hour--not minutes.)

K. Farmer Brown

Farmer Brown is going to use all 70 acres of his farm to plant corn or wheat. Based on past history, he gets 110 bushels of corn per acre and 30 bushels of wheat per acre. He expects to earn $1.30 per bushel net profit on corn and $2.00 per bushel net profit on wheat. The total net profit for Farmer Brown is based on the number of acres he plants in corn (C) and the number of acres he plants in wheat (W).

> Placing units in a calculation is a double check to see that the formula is used correctly. Verify that the units simplify and the final answer is in Dollars. The verification is part of "dimensional analysis."

How much net profit should he expect on 48 acres of corn and 22 acres of wheat?

The formula is

$$P = (110 \tfrac{bushels}{acres})(\tfrac{\$1.30}{bushel})(C) + (30 \tfrac{bushels}{acres})(\tfrac{\$2.00}{bushel})(W)$$

This simplifies to

$$P = (\tfrac{\$143.00}{acre})(C) + (\tfrac{\$60.00}{acre})(W)$$

Simplifying, we find

$$P = (\tfrac{\$143.00}{acre})(48 \ acres) + (\tfrac{\$60.00}{acre})(22 \ acres)$$

$$P = \$6864 + \$1320$$

$$P = \$8184$$

```
143*48+60*22
        8184.000000
■
```

Answer: *The profit for Farmer Brown on 48 acres of corn and 22 acres of wheat is $8184.*

L. Time is Money

The number e is a mathematical constant found in many applications. It is approximately equal to 2.718281828459. It can be found in applications involving electrical circuits, population growth and in the mathematics of finance. The Present Value (PV) or Future Value (FV) can be calculated given an annual interest Rate (r) and the number of years (t) according to the following formulas.

$$FV = PV (e^{\,jt}) \qquad\qquad PV = FV (e^{\,-jt})$$

In 1970, before the first gasoline crunch, it cost approximately $3.00 to put 10 gallons of gasoline in your car. Find the value of that $3.00 in 1997 using an annual interest rate of 6%.

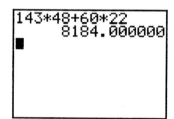

Calculators often have a button to perform powers of e so you don't have to use the decimal approximation.

$e^{5.6}$ – 5.6 e^x = 270.4264

The Future Value of $3 over 26 years at 6% is

$$FV = \$3.00 \ (e^{.06 \cdot 26})$$

$$FV = \$3.00 \ (e^{1.56})$$

$$FV = \$3.00 \cdot 4.75882$$

$$FV = \$14.28$$

Answer: *The price of 10 gallons of gasoline in 1997, based on the 1970 costs, should be $14.28 or about $1.43 per gallon.*

A family decides it wants to have $100,000 in the bank in 15 years for several purposes. If it assumes an 8% growth rate, how much money must be placed in the bank today in order to grow to $100,000 in 15 years?

The Present Value of $100,000 15 years from now using an annual rate of 8% is

$$PV = \$100000 \ (e^{-.08 \cdot 15})$$

$$PV = \$100000 \ (e^{-1.2})$$

$$PV = \$100000 \ (.301194212)$$

$$PV = \$30119.42$$

Answer: *The family would need to place $30,119.42 in the bank at 8% in order to have $100,000 in 15 years.*

M.   Animal Farm

The previous examples have demonstrated several types of models. The Straight-line Depreciation was an example of a linear model and the Time Is Money was an example of an exponential model. The following example will consider the population of an animal in a closed environment. This environment could be a cage, an island, a pond, a garden, a continent, or a planet. Because the resources are limited, the population can not continue to grow (that rules out the exponential model). However, the population will grow faster initially as it gets more numerous (that rules out the linear model). The type of model that fits this situation is a **logistic growth** model. The logistic growth model is an example of a *recursive formula*. A recursive formula is one in which every element after the first can be obtained from the preceding one.

```
3(e^(.06*26))
        14.27646374
100000(e^(-.08*1
5))
        30119.42119
```

When calculations involve a negative exponent, care must be taken when using a calculator. It is advisable to have some idea (an estimate) as to what to expect as the answer.

When a positive number greater than one is raised to a power which is less than zero, the result is a number between 0 and 1.

That is, if $a > 1$ and $p < 0$, then
$$0 < a^p < 1$$

Examples:
$$2^{-3} = 1/2^3 = 1/8$$

$$5^{-2} = 1/5^2 = 1/25$$

Therefore, when you multiply a number by a positive number greater than one which is raised to a negative power the result will be a number smaller than the original number.

So,
$$1000(5^{-2})$$
$$= 1000 \ (1/25)$$
$$= 40$$

A simple example of a recursive formula is an *arithmetic sequence*. The definition of an arithmetic sequence can be stated symbolically by giving the value of the first element a_1, the number of elements n, and the formula

$$a_{n+1} = a_n + d$$

This formula indicates that the $n + 1$ *st* term is d more than the *nth* term - no matter what n is. So if $a_3 = 5$ and $d = 3$, then $a_4 = a_3 + d = 5 + 3 = 8$. The next term can also be found in this way: $a_5 = a_4 + d = 8 + 3 = 11$, and so on.

The geometric sequence is another example of a recursive formula. The formula for a geometric sequence is

$$a_n = a_1 \, r^{n-1}$$

In this formula the *nth* term is dependent on the first term, a_1 and a value r. If $a_1 = 200$ and $r = 1.06$, then the *10 th* term is

$$
\begin{aligned}
a_{10} &= a_1 \, r^{n-1} \\
&= 200 \, (1.06)^9 \\
&= 200 \, (1.689478959) \\
&= 337.90
\end{aligned}
$$

Arithmetic and geometric sequences are often seen as lists. Consider the following examples. Determine whether the list is an example of an arithmetic sequence or a geometric sequence. If it is arithmetic, determine its constant difference, d. If it is geometric, determine its constant ratio, r. If it is neither, so state.

No. Of Terms	1	2	3	4	5	6
Term	5	7	9	11	13	15

No. Of Terms	1	2	3	4	5	6
Term	2	4	8	16	32	64

No. Of Terms	1	2	3	4	5	6
Term	2	3	5	8	13	21

No. Of Terms	1	2	3	4	5	6
Term	5	15	45	135	305	915

$a_{n+1} = a_n + 3$	
Term	Value
1	5
2	8
3	11
4	14
5	17

$a_n = a_1 \, 3^{n-1}$	
Term	Value
1	2
2	6
3	18
4	54
5	72

Subscripts have the affect of producing a *different variable* using the same letter. That is, the variable b_1 is different from the variable b_2. It is customary for b_1 to represent the **first** value of the variable b, b_2 the second, and so on. b_n, then, represents the *nth* variable.

The growth of a population within the confines of a finite space, or *habitat*, depends on the size of the habitat (which has something to do with the resources available to the population) and on the population's ability to reproduce. The formula for a logistic growth model is

$$p_{n+1} = r\,(1 - p_n)\,p_n$$

where p_n is the percent of the greatest population the habitat will support. So if $p_n = .25$, then the population is at 25% of the greatest number possible. Thus, p_n is between 0 and 1 $[0 \le p_n \le 1]$. The variable r is dependent on the animal being studied and is between 0 and 4 $[0 \le r \le 4]$.

This model also lends itself well to an analysis using a **graph**. The graph shows how p_n (y-axis) changes over time (x-axis). For these values of **p** and **r**, this population stabilizes. The next two examples (shown on the next page) are graphed using a spreadsheet to perform the calculations. The first graph shows a population which is stable with about a 10% change seasonally. The second graph demonstrates conditions which cause a *chaotic* change in the population.

Consider the number of wild horses on an offshore island. Show the change in the population over several *seasons* (the length of a season depends on the animal being studied) if we begin with $r = 2$ and $p_1 = .3$

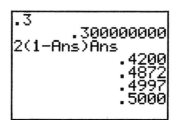

$$p_2 = 2\,(1 - .3)\,.3 = .42$$

$$P_3 = 2\,(1 - .42)\,.42 = .4872$$

The next several calculations are shown here. For brevity, only the results are shown. You may want to verify some of the results.

$$p_4 = .49967 \quad p_5 = .4999998 \quad p_6 = .5$$

Each calculation thereafter results in .5 which means that the population is holding at 50% of the maximum value.

Find the change in the population of fish in a pond if

$$r = 2.5 \quad and \quad p_1 = .15$$

and graph the population over 6 seasons.

$$p_1 = .15$$

$$p_2 = 2.5\,(1 - .15)\,.15 = .31875$$

$$p_3 = 2.5\,(1 - .31875)\,.31875 = .54287$$

$$p_4 = 2.5\,(1 - .54287)\,.54287 = .62041$$

$$p_5 = 2.5\,(1 - .62041)\,.62041 = .58876$$

$$p_6 = 2.5\,(1 - .58876)\,.58876 = .60531$$

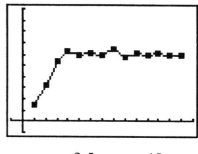

r = 2.5, p = .15

Logistic Growth Models								
P1	0.15	Season	Pn					
r =	3	1	0.15					
		2	0.3825					
		3	0.708581					
		4	0.619482					
		5	0.707172					
		6	0.621239					
		7	0.705904					
		8	0.622811					
		9	0.704752					
		10	0.62423					
		11	0.703701					
		12	0.625518					
		13	0.702736					
P1	0.35	Season	Pn					
r =	3.8	1	0.35					
		2	0.8645					
		3	0.445131					
		4	0.93856					
		5	0.219128					
		6	0.650222					
		7	0.864246					
		8	0.445834					
		9	0.938851					
		10	0.218158					
		11	0.648147					
		12	0.8666					
		13	0.439298					

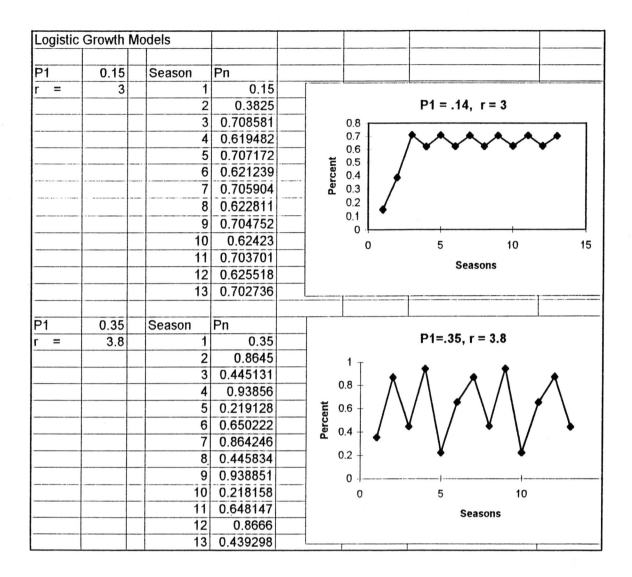

1.3 Investigating the Models

Objectives of this section. The student will be able to
- *use the formulas in section 1 to compute the values requested*
- *use the formulas to evaluate characteristics of the model*
- *use the formulas to produce numerical tables*
- *use the formulas to produce graphs*

The problems in this section refer to the applications defined in section 1.1.

☺A. The Tax Rate Table
1. Find the Tax Owed if the Net Taxable Income is $8,500.
2. Find the model for the Tax Owed if the NTI is over $10000. Use this model to find T if NTI = $15,500
3. Find the Tax Owed if the Net Taxable Income is $5,999, $6,000, $6,001.
4. Problem 3 points out an interesting phenomenon in this model. The amount of Tax Owed doesn't "jump" as we change from one rate to another. Using a rate of $140 + *5%* on $6000 produces the same Tax Owed as using the rate of $240 + *6%* on $6000.

   Suppose, in a subsequent year, the Tax Rate table used different percentages and a different interval. Find the values *xxx* that produce a smooth change as was noted above in the 1985 Tax Table.

TAX RATES

If Net Taxable Income (line 15) is

over	but not over	The Tax is
-0-	$12,750	6% of net taxable income
$12750	$60000	$xxx + 7% of amt. over $12750
$60000	---	$xxx + 7.75% of amt. over $60000

NC Tax Rate Table (1995)

☺B. Base Salary Plus Commission
1. Find Jake's monthly Salary if his Total Sales for a month were $30,525.
2. Find Jake's monthly Salary if his Total Sales for a month were $9,525.
3. What Total Sales must Jake produce in order to earn a Salary of $3,000 per month?
4. If Jake were sure he could get a least $15,000 in sales each month, he would not have had to ask for the $1,000 base salary. How much does Jake "lose" each month that he exceeds $15,000 in Total Sales because of the base salary clause?
5. Elroy, Jake's brother, is also negotiating a contract to work for Blue's Copiers. If Elroy gets a base salary of $2,000 plus 8% commission on sales, at what Total Sales figure should Elroy begin to earn commission? Construct a model for Elroy's Salary and find his Salary on Total Sales of $27,205.

C.   The Winner's Share

1.   Find the Individual Prize Money if the Number of Winners were five and 1444 tickets were sold.

2.   Find IPM if W = 6 and TS = 1444.

3.   If TS stays the same and W increases, what will happen to IMP? Why?

4.   What is the smallest number of tickets that can be sold and still provide a prize?

5.   What is the smallest number of tickets that can be sold so that the IPM for 4 winners is at least $100?

6.   The following spreadsheet shows the relationship between the IPM and the TS and W as both variables change. Why do the numbers in each row increase as you move from left to right? Why do the numbers in each column decrease as you move down? Can you identify any other patterns? Can you explain them?

Number of Tickets Sold

No. of Winners		750	1000	1250	1500	1750	2000
	1	500.00	1,000.00	1,500.00	2,000.00	2,500.00	3,000.00
	2	250.00	500.00	750.00	1,000.00	1,250.00	1,500.00
	3	166.67	333.33	500.00	666.67	833.33	1,000.00
	4	125.00	250.00	375.00	500.00	625.00	750.00
	5	100.00	200.00	300.00	400.00	500.00	600.00
	6	83.33	166.67	250.00	333.33	416.67	500.00

D. Future Value

1. Find the Future Value of $5000 invested at 6% compounded monthly over five years.

2. Find a model to find the F if r = 5% and t = 72 months.

3. Use the model in (2) to show that F = $6745.09 for $5000 invested monthly at 5% for six years.

4. The table shows the total interest earned on $10,000 at 6% compounded monthly. Verify at least one of the columns of figures.

   The third row shows the *average monthly interest* over the five years. Explain why the average monthly interest seems to decrease to about $50 as the number of months the investment is held decreases?

	6 Months	12 Months	18 Months	24 Months	30 Months	36 Months	42 Months	48 Months
Total Interest	303.78	616.78	939.29	1271.60	1614.00	1966.81	2330.33	2704.89
Average Monthly Interest	50.63	51.40	52.18	52.98	53.8	54.63	55.48	56.35

☺E. The New Room

1. Find the total cost for the two projects if the new room is 18 ft. 3 in. × 16 ft. 9 in.

2. The dimensions of the new addition are shown to the right. Calculate the total cost of the two projects for this addition.

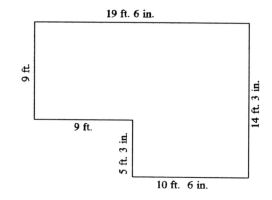

F. The Bullet's Path

1. A bullet was fired at a man standing on a building. The bullet grazed the man's head. In order to locate the bullet more easily, at what height would the bullet have hit the building?

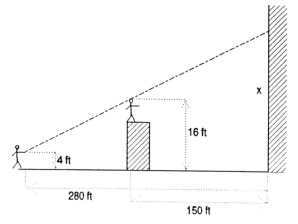

2. How much shorter is it from point A to point B using the direct route?

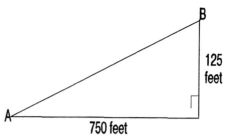

125 feet

A

750 feet

B

3. A man-hunt was planned for the area between three roads as shown in the figure. If each man can cover an area of 250 square feet in an hour, how many men will it take to cover the area in one hour?

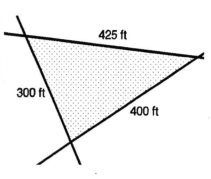

425 ft

300 ft

400 ft

G. Straight Line Depreciation

1. Find the Annual Depreciation on an asset with an Initial Value of $38,000 and a Scrap Value of $2,500 over 10 years.

2. In order to visualize the relationship between two variables, it is useful to be able to plot ordered pairs. (You have probably had experience graphing points on an coordinate axis using points labeled (x,y).)

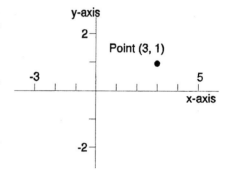

y-axis

2

Point (3, 1)

-3

5

x-axis

-2

The following graph is a picture of the relationship between the number of years and the book value of an asset that had an Initial Value of $13,050, a Scrap Value of $1000 and a life of 7 years.

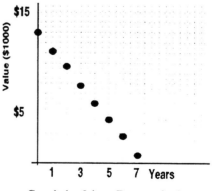

$15

Value ($1000)

$5

1   3   5   7 Years

Straight-Line Depreciation

Use the grid to the right to plot the same type of graph for the information given in problem #1 of this group ($3,8000 Initial Value, $2,500 Scrap Value, life of 10 years).

Before plotting the points, it will be necessary to determine a scale for the x- and y-axis. For example, the x-axis can be scaled by letting each mark equal 1 year and the y-axis may be scaled by letting each mark equal $2,000.

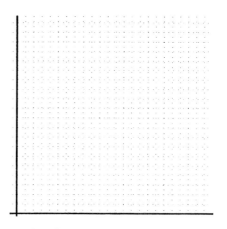

Straight-Line Depreciation

H.   Monthly Payments

1.   Find Betty's monthly payment if she needs to borrow $9600 at an APR of 7.98% for 48 months.

2.   Find Betty's monthly payment if she needs to borrow $9600 at an APR of 4.75% for 48 months.

3.   Which situation yields a lower monthly payment on a loan of $5500:

Bank 1:   Offers an APR of 5% over 60 months

Bank 2:   Offers an APR of 4.5% over 48 months

Which would you choose and why?

4.   The spreadsheet below shows the Monthly Payment required to amortize a $10,000 loan. Why do the numbers in each row decrease from left to right? Why do the numbers in each column increase moving down? If the amount of time to pay off a loan is doubled, does the payment get cut in half? Why? Do you think the payment for a loan at 12% is twice the payment on the same loan at 6%? Explain.

Monthly Payment Required on $10,000

Number of Months

		24	30	36	42	48
	.06	443.39	359.79	304.32	264.59	234.84
Annual Rate	.07	447.74	364.38	308.80	269.15	239.49
	.08	452.47	368.87	313.40	273.77	244.16

I.   True Cost

1.   Find the Cost per Mile for Lee if L = 2000, Y = 36,000 miles and all other costs remain the same.

2.   Find the simplified model in terms of L and Y if Tires are $400 per 40,000 miles, Insurance is $1400 per year, Expenses are $750 per year, and all other expenses remain the same.

3.   Gasoline is shown at $.06 per mile. The following computation is based on gasoline at $1.20 per gallon and with a car getting 20 miles per gallon:

$$Cost\ of\ Gas\ per\ Mile = \frac{\$1.20\ per\ gallon}{20\ miles\ per\ gallon}$$

$$C = \frac{\$1.20}{1\ gallon} \div \frac{20\ miles}{gallon}$$

$$C = \frac{\$1.20}{1\ gallon} \times \frac{1\ gallon}{20\ miles}$$

$$C = \frac{\$.06}{mile}$$

Find the cost of gasoline per mile if a car gets 16 miles per gallon and gasoline costs $1.099.

☺J.   Distance Rate and Time

1.   Find the Time (in minutes) it takes to travel 45 miles at an average rate of 55 miles per hour.

2.   Find the Average Rate (in feet per second) necessary to travel 1000 feet in 12 seconds.

3.   Find the Distance traveled at 44 feet per second for 1.5 minutes.

4.   If a person traveling from A to B averages 3 feet per second going the direct route, how fast must a person average taking the long route in order to make the trip from A to B in the same amount of time?

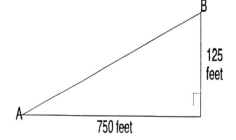

☺K. Farmer Brown

    1.   How much profit should be expected on 30 acres of wheat and 40 acres of corn?

    2.   Over the next few years, due to export limitations imposed on farmers and changing weather conditions, the profit on wheat goes to $1.75 per bushel and the profit on corn rises to $1.45 per bushel. If the farmer now averages 100 bushels of corn per acre and 35 bushels of wheat per acre, find a model to calculate the Profit.

            Using the previous scenario, find the Profit if Farmer Brown plants 22 acres of corn and 48 acres of wheat.

L. Time is Money

    1.   Find the value of an item in 25 years if the PV = $.25 and we assume an interest rate of 6%.

    2.   Find the Present Value of $50,000 in 20 years assuming an interest rate of 10%.

M. Animal Farm

    1.   Make the calculations and show a graph of the population if $r = 3$ and $p_1 = .3$.

    2.   Make the calculations and show a graph of the population if $r = 1.5$ and $p_1 = .25$.

Chapter 2   Mathematical Modeling

In Chapter One you answered questions and performed calculations on mathematical formulas (or models) that were provided. These formulas allowed you to investigate complex physical phenomenon using mathematical tools. When formulas are not available, mathematical models must be *developed* to describe the real world situation. The development of a good model can be very difficult. Often, there can be no one model that is absolutely correct. The task then is to find the *best* model with the available resources.

The following activities will guide you through the experience of finding a model.

Activity #1   **Two-Digit Multiplication**

One of the objectives of the early grades is for students to learn the multiplication table for single digit factors. Some probably memorized the table up to 20. Normally, two-digit multiplication is a task performed with pencil and paper (if not a calculator). The following activity will develop a model that will allow two-digit multiplication to be performed mentally with the aid of a model and will illustrate the process.

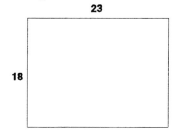

Figure 1

Consider the product of 23 and 18. First, draw a rectangle (Figure 1) to represent an area that is 23 × 18. (Recall that the area of a rectangle is the *length* times the *width*, so 23 × 18 is the area of this rectangle.)

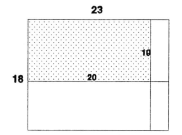

Figure 2

Next (Figure 2), divide the rectangle into one large rectangle and three smaller ones. The dimensions of the largest rectangle within the original rectangle will be the largest multiple of 10 which is less than the original length and width. In this case the large rectangle will be 10 × 20.

The dimensions of the other rectangles can now be found. The dimensions of each rectangle are used to find the area of each rectangle constructed inside the original rectangle. See Figure 3. The area of the original rectangle (and the product of the two factors) is then found by adding the four areas of the interior rectangles.

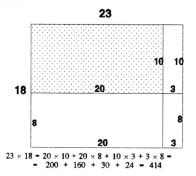

23 × 18 = 20 × 10 + 20 × 8 + 10 × 3 + 3 × 8 =
   = 200 + 160 + 30 + 24 = 414

Figure 3

This process is demonstrated again in the figure below. Figure 4 demonstrates the product of 36 × 24 = 864.

After constructing several of these models, can you perform the arithmetic mentally?

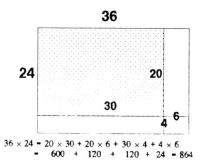

36 × 24 = 20 × 30 + 20 × 6 + 30 × 4 + 4 × 6
   = 600 + 120 + 120 + 24 = 864

Figure 4

	An athlete signs a contract for $1 million per year.

If you read much about the world today you will encounter large numbers, such as the salary of CEOs or of athletes, the trade deficit, or the national debt. Most people do not really understand how *large* large numbers are! This activity will help us develop an appreciation of large numbers by developing a model to use for predicting.

The congress talks about the budget in increments of $100 million.

To help illustrate large numbers, consider 1,000,000 pieces of regular 20-pound typing paper. How high would 1,000,000 sheets of 20-pound paper reach?

Before reading on, guess how high 1 million sheets of paper will reach!

In order to find the model for this activity, you will have to create some data. Find a ream of 20-pound paper and measure the actual height of just the 500 sheets of paper.

The height of 500 sheets is about 2 inches. Now, use a proportion to create the model:

$$\frac{Height}{Number\ of\ Sheets\ of\ Paper} = \frac{2\ inches}{500\ sheets\ of\ paper}$$

$$Height = No.\ of\ Sheets\ of\ Paper \left(\frac{2\ inches}{500\ sheets\ of\ paper} \right)$$

Substituting 1,000,000 sheets of paper into this model, we find

$$Height = 1,000,000 \left(\frac{2\ inches}{500\ sheets\ of\ paper} \right)$$

$$Height = 4,000\ inches$$

Try doing these calculations in metric units. 500 sheets of 20-pound paper is .05 meters.

Four thousand inches is over 333 feet or about the length of a football field and one end zone. Can you think of a structure that would be about this tall? One million is a large number, especially when we consider 1000 sheets of paper would only be 4 inches high, and 10,000 sheets of paper would only be slightly more than a yard. If someone gave you $10,000 to spend on a vacation that would make you happy. How much happier would you be if they gave you $1,000,000.

As big as one million is, larger numbers are in the news daily. Complete the table for the numbers shown.

	Large Numbers						
Number of Sheets of Paper	500	1000	10,000	1,000,000	10,000,000	100,000,000	1,000,000,000
Height (use the correct units)							
Compare this height to something real	little finger	coffee cup	door knob				

This activity will require you to acquire some information from other sources and then find a model that can be used to predict the length of a year for each planet based on the earth's year. (A year as being defined by the time it takes that planet to revolve around the sun.) We know that the earth is about 93.0 million miles from the sun. If Mars is 141.6 million miles from the sun, how long is a Mars Year?

The model will be derived from proportions.

$$\frac{A\ Year}{365.25\ Earth\ Days} = \frac{Distance\ from\ the\ sun}{93,000,000\ miles}$$

$$A\ Year = 365.25\ Earth\ Days\left(\frac{Dist.\ from\ the\ sun}{93,000,000\ miles}\right)$$

Now substitute the distance Mars is from the sun.

$$A\ Year = 365.25\ Earth\ Days\left(\frac{141,600,000\ miles}{93,000,000\ miles}\right)$$

$$= 556.12\ Days$$

The actual Mars Year is 687 Earth Days. While our model didn't come really close, it still might provide estimates for the other planets. Complete the table to the right for each of the other planets.

For which planet was our estimate the best? the worst? Why?

The distance from the sun should have something to do with the length of time it takes to revolve around the sun since the circumference of a circle is dependent of the radius of the circle:

$$C = 2\pi r$$

Planet	Distance from the sun (millions of miles)	Length of one year (in Earth days)
Mercury	36.0	
Venus	67.2	
Earth	93.0	365
Mars	141.6	556
Jupiter		
Saturn		
Uranus		
Neptune		
Pluto		

The ability to count the number of ways something could or might happen is often required. In this activity, you will utilize patterns to construct a numerical model that can be used to count the number of games it takes for every team in a league to play every other team. The Knightdale 11-12 year-old baseball league is composed of five teams. How many games will it take for each of the five teams to play each of the other teams one time?

We will use a method often used in investigations--start with a simpler problem. Suppose the number of teams is only two, then three, then four, then five. Hopefully, the answers will yield enough information to enable us to continue without much work. The figures to the right illustrate the work described below and summarized in the table.

Figure 5 shows that for only two teams it takes one game. (A dot represents a team and a line connecting two dots represents a game.) Figure 6 shows that for three teams it takes 3 games. Figure 7 shows that for four teams it takes six games.

Use Figure 8 to count the number of games needed for five teams. Organize the information you have found in the table below.

With this information can you complete the second row of the table? Probably not--at least not with confidence in the accuracy of our prediction! The pattern is not obvious. So that is why the third row is there.

Use the third row of the table to represent the change in the number of games it takes when you add one more team. Write ΔG in the first column for the third row. With this information, you can now complete the table.

Two teams

Figure 52

Three teams

Figure 53

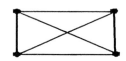

Four teams

Figure 54

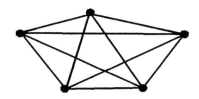

Five teams

Figure 55

Number of Teams	2	3	4	5	6	7	8
Number of Games Needed	1	3	6				
		2					

If not checked, a contagious disease can spread quickly through a population. The problem with the "mad-cow" disease in England in the mid-90's illustrates the fact that measures must be taken quickly to check the spread of a disease--whether in cows or in people! The following activity will model the spread of a contagious disease in a herd of 1,000 cows. The activity could as easily have used a city of 1000 persons.

On *Week 0* **ten** infectious cows are introduced to the population. Each week thereafter, an **additional 20%** of the population is infected with the disease. Complete the table to the right using this information.

A graph will be employed in order to make predictions about the spread of the disease. The graph shows the spread of the disease over 10 weeks. Check your numbers in the table against those in the graph.

About how many weeks (or days) did it take for 50% of the population to be infected?

About how many weeks (or days) did it take for 75% of the population to be infected?

About how many weeks (or days) did it take for 90% of the population to be infected?

Why is the curve sloping down as the number of weeks increase? Explain.

Will the entire population ever be infected? Explain.

What problems do you see with this model? Explain.

Find a source to look up some facts about sixteenth century history. Find out the name of the disease the Spanish troops introduced that lead to the fall of Montezuma.

20% Infection Rate

WEEK	Number of Healthy Cows	Number of Cows Newly Infected	Total Number of Cows Infected
0	1000	10	10
1	990	198	208
2	792	158	366
3	634	127	493
4	507	101	594
5	406		
6	325		
7			
8			
9			
10			

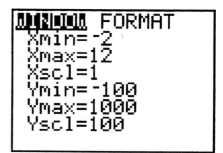

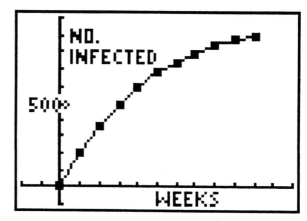

☺**Activity 1**

1. Compute   a.   76 × 57

                 b.   42 × 84

2. What would the model for three-digit multiplication look like? Construct an example?

☺**Activity 2**

1. Find a good approximation for the height of 1 billion pennies.

2. Find a good approximation for the area that 1 million one-dollar bills would cover if laid next to each other. What percent of the earth's surface is that?

Activity 3

1. Locate a sink with a water faucet that can be used to fill up a gallon milk jug. If the cold-water faucet is turned on completely, how long will it take for 100 pounds of water to run out?

2. A pool has dimensions as shown. The density of water is

$$\frac{1 \ kg.}{m^3}$$

Calculate the amount of time the faucet used in problem 1 will take to fill a pool with the dimensions shown here.

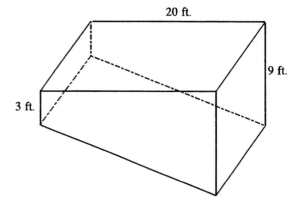

Activity 3

☺**Activity 4**

1. A polygon is a figure constructed of straight sides that connect at a vertex. A rectangle is a polygon with four sides and a pentagon is a polygon with five sides. A diagonal is a line segment that joins two vertices that are not connected by sides. Find a model to predict the number of diagonals for an *n-sided* polygon.

         Rectangle   **Pentagon**

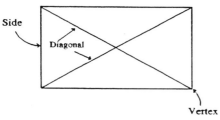

Activity 4 #1

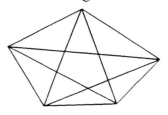

Activity 4 #1

2. Find a model to predict the number of squares in figures such as the ones to the right.

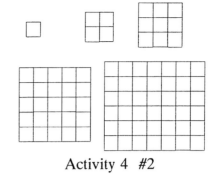

Activity 4 #2

3. Some of the squares that make up a larger square are shaded. If the pattern continues, find a model to predict the ratio of the number of shaded squares to the number of non-shaded squares in a large square.

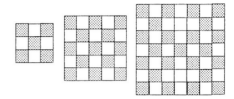

Activity 4 #3

4.   a)   Find a model to predict the number of small cubes in each large cube.
    b)   Find a model to predict the number of small cubes in each large cube if **one** of the rows of cubes has been removed.

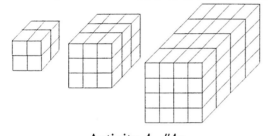

Activity 4 #4a   Activity 4 #4b

Activity 5

1. Repeat the model with the following changes:
   - there are 5000 people
   - the infection rate is 10%
   - initially only 2 people are infected.

2. Accumulate a large handful of pennies. (Count how many you start with and define that number as the initial population.) Use the following model to *simulate* a condition like the problem described in Activity 5. Each time you drop the pennies represents one week. Every penny that comes up tails has been *infected* and will be removed. Keep track of the number of pennies **left** after each "week." The number of pennies remaining simulates the people who are not infected. Graph this model over several weeks and compare this simulation with the model in Activity #5.

Every day attempts are made to educate, influence, bewilder, impassion, and implore using statistics. In order to function effectively in this world, consumer/citizens must be able to understand the information inundating us and understand how the statistics were derived. Statistics is often called the "science of data." This science will enable you to become effective producers and users of data.

3.1 Sampling

Objectives of this section: The student will be able to
- *define population, sample, bias and random sample*
- *be able to apply thee previous definitions to applications which involve them*
- *design an experiment to sample a large population without bias*

A local afternoon talk show took a poll of its listeners. The poll was taken to determine what people thought of an expenditure of over $50,000 for an "art tower" on a busy highway. The host reported that over 88% of his listeners disagreed with the expenditure. Furthermore he said, based on his poll, that most people in the city disagreed with the expenditure.

The
Art Tower
Question

Populations and Samples

In order to use statistics as evidence, the data must be collected properly. The collection of data can not be trivialized. It is important to observe recognized scientific procedures to ensure that the data are accurate and unbiased.

Often conclusions about a entire group are drawn from the information taken from only part of the group--a *sample*. Phrases "like 67% of all doctors recommend," or "two-thirds of our owners believe" are usually statements about a *sample* taken from a *population*.

A **population** is a set that describes some phenomenon of interest to the statistician.

3.11

A **sample** is a subset of the population.

3.12

---

a. Suppose a population consisted of 4,352 people. If a sample of 300 people were taken from this population, what percentage of the population would have been sampled?

b. 642,244 people make up a population. If a sample of 1.5% is to be taken, how many people will be sampled?

The State of North Carolina considered changes to the driver's license all drivers must carry. Before making the final decision, the committee assigned the task of making the changes decided to get input from the people who carry driver's licenses. Since it was not feasible to ask everyone, the committee decided to survey the people who came into one of its renewal offices on Monday, September 14. Based on their input, no changes were made.

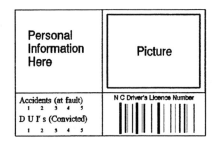

The Proposal

Questions
1. Describe the population.

2. Describe the sample taken.

3. About what percentage of the population do you think was sampled?

A population need not consist of people. Populations may be made up of objects such as electric starters produced at Square D or animals such as the wild horses on Bald Head Island. For many reasons, we often only gather information about a few items selected from a population. The part of the population used to draw conclusions about the whole is the *sample*.

Scenario #1

A company in the RTP produces fuses. These fuses are suppose to break the flow of current if the current reaches a certain level. The company must be sure that the fuses do what they are intended to do. To do so the company must perform tests on the fuses to see how they perform. Discuss why it be would be appropriate to use a sample instead of testing the entire population.

Scenario #2

A company considering relocating to the Triad area wanted to know the average educational level of the people in that area. Why would a sample be considered?

Bias

Sampling must always be protected from *bias*. Whenever we allow our own convenience to dictate how a sample is obtained, we run the risk of misrepresenting the facts about the population. Bias is an error that occurs because of design, not luck.

> **Bias** is the difference between the results obtained by sampling and the truth about the whole population.

3.13

Applying the Concept

A company was employed to gather facts about Wake County citizens to use in an ad campaign to attract industry. It decided to set up a booth at Crabtree Valley Mall, inside an elite men's clothing store. Discuss the bias that might be present in the data the company uses.

In order to eliminate bias from a sample we must ensure that our sample is representative of the population. The most common way to satisfy this requirement is to select the sample in such a way that every different sample of size *n* has an equal probability (or chance) of being selected. This procedure is called a *random sample*.

> A **random sample** of size *n* is a sample drawn in such a way that every possible sample of *n* members of the population has the same chance of being selected in the sample chosen.

3.14

What methods can be used to produce a random sample? If it is possible to assign every member of the population a number, then several means of obtaining a representative sample are possible. For example, if the population to be studied consisted of the faculty at Wake Technical Community College, then we could assign each faculty member a number--say 1 to 356. Any number of procedures could them be employed.

For example.

o   Take three containers and into the first, place the digits 0,1,2, and 3 written on white poker chips, and into the other two, place ten chips with the digits 0 through 9 written on them. By selecting one chip from each container, a random sample could be chosen.

o   The sample could be selected by using a computer program that would select each faculty member randomly using a random number generator.

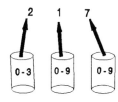

Table of Random Numbers

Random numbers are often used for samples. A <u>table of random numbers</u> can be found in most statistics texts or in book of tables in the library. This table is simply a lot of numbers randomly chosen and written in what seems to be an endless list. The numbers are used to produce representative samples. For the previous example, look at the first three (or any three) digits of the random numbers from a table. If the number was greater than 356, it would not be used.

		Table of Random Numbers			
940712	598759	172579	763210	091193	524362
104382	287591	659265	938473	105692	350982
361096	211604	358393	019103	843882	358233
928133	390247	439858	385022	947490	109811
742010	284064	210143	382011	438201	395011
438201	298419	940904	984820	438290	345977
438206	507598	472782	447109	856016	109813
510925	609256	209560	923847	853103	678131
484902	113833	028522	424242	928100	789100
525929	569922	103335	150326	114141	018513
875010	298234	913954	133119	290820	019071
295921	231515	371820	924901	582171	736191
932952	332234	710308	838652	019313	186134
029293	950212	104158	019492	102983	937219
457298	651063	096350	961029	949551	104376
190202	378252	356221	646234	018751	193479
835101	883521	427298	958256	645677	987361

Find a twelve-person sample of the faculty at Wake Technical Community College using the portion of a random number table.

Note: Before beginning, decide which three digits in the random number to use. For example, you may choose to use the first three digits

1 2 7 3 7 8

or three in the middle

1 1 5 7 8 8.

If it is not possible to assign each member of the population a number, then several steps must be taken to ensure that a representative sample is obtained.

Samples

When polls are taken to predict elections or to obtain the perception of the population, samples of the entire population of the U.S. are taken. The pollster may determine that 1200 households are to be sampled. Those 1200 households must be selected so that some are in rural areas, some in urban areas, some in the east, and some in the west. Suppose that the pollster decides to receive input from 5 major cities (population > 1,000,000), 10 cities (200,000 < p < 1,000,000), 20 towns (p < 200,000) and 30 rural areas. Raleigh has been selected as one of the cities from which 10 households will be polled. How will the ten households be selected? One way to select the households would be based on geography; two would be selected from each of the five regions of the city: central, eastern, western, northern, and southern. The two households from each region could then be selected at random from a tax list kept in the county's tax office.

Group Work

Divide the city of Raleigh (or your home town) into five regions from which the two households will be selected. Is there a better way to select 10 households from the city of Raleigh who will participate in this poll? Describe another way to divide this area in order to select the 10 households.

1. Write a paragraph about the findings of your group work for scenarios #1 and #2 in today's class.

☺2. Write a short paragraph to compare and contrast a sample and a population. Discuss when each might be used in statistical study.

3. Write a short paragraph to explain how bias could be used to misrepresent a population for the personal good of an individual or company.

☺4. Find a paperback novel or short story. We will use the table of random numbers to select 20 words at random from different pages in the book. Decide first how you will select the page to use and then the how you will select the number of the word on that page that will be used. Then, use the table of random numbers to generate 20 pairs of numbers. What is the average word length of your 20 words? Compare your answer with others in the class.

5. Find several examples that use statistics or of polls in recent newspaper articles. Cut them out and bring them to class.

6. Discuss the problems with the conclusion the host arrived at in the vignette at the beginning of the section. What do you think *voluntary selection* means? Give an example.

3.2   Describing Data

Objectives of this section: The student will be able to

- *display data graphically using either pie charts or histograms.*

- *deduce information from pie charts and histograms.*

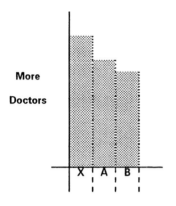

More doctors
recommend
Preparation X
over any other brand.

Misleading
Graphic

Graphics

While some people would rather look at all the numbers, others would like for the data to be condensed into a "picture" or "graphic." USA Today is notorious for using many graphs, pie charts and other graphics to display statistics. These graphical methods for describing data essentially show how many observations fall into each category. The categories that are chosen to classify the data are called *classes*. Once the classes have been determined, each piece of data is placed into one of the classes. The number of observations falling into a class is the *frequency*.

> The *frequency* for a *class* is the number of observations falling in that class.

3.21

Pie Charts

Pie charts usually show the proportions or percentages of the total number of measurements falling into each class. Although these graphics can be constructed by hand, you will probably rely on computer software to produce the graphs for presentations.

a)   Use the information provided in the pie chart to answer the questions
below.

A retail outlet in a mall recorded $244,025 in sales this past month. 78%
of the sales in dollars were charged to major credit cards. About how
many dollars in sales were charged to Master Card? to Discover?

If Visa charged this customer .25% to process the sales to Visa, how much
did this outlet pay Visa for their services?

Does the pie chart indicate that more people own the Visa Card than own
Master Card? Explain.

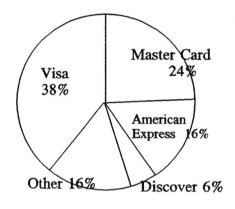

**Major Charge Card's Share of
Dollars Charged in 1993**

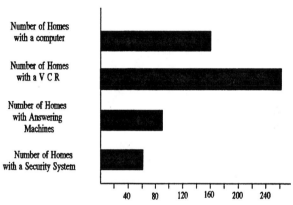

From a Survey of 400 Homes

b)   Use the information provided in the histogram to answer the
questions below.

About what percentage of the homes surveyed owned a VCR? a
computer?

Does this graph indicate how many homes own a VCR and an
answering machine?

Measuring Angles

In order to construct accurate pie charts you must be able to measure angles. One way to measure angles is with degrees. In one complete sweep of a circle, there are 360°.

Measuring Angles
One Complete Sweep

Therefore, any portion of a sweep will be the same portion of 360°; and more importantly, any portion of 360° will be that portion of a complete sweep. If one portion of your pie chart is to be 20% of the entire circle, then the measure of the angle that would be used would be 20% of 360° or 72°. If a portion of the pie chart is to represent 75% of the entire circle, then the measure of the angle would be 75% of 360° or 270°. When constructing pie charts by hand, you may use a protractor or carefully estimate the measure of the angles.

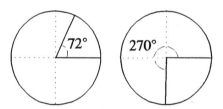

Angles in Degrees

Group Work

One side of an angle is drawn in each figure to the right. Complete the angle by drawing the other side to match the measure indicated.

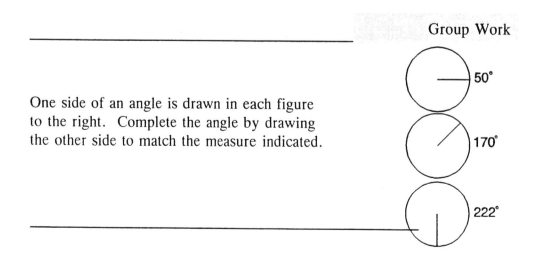

Histograms

The classes for histograms are usually divided in one of two ways.

If the data is numerical (time is a good example), then the data is usually separated into equal intervals. For example, the data 4.2, 4.3, 4.4, 4.4, ...5.2, 5.4, 5.7 ranges from 4.2 to 5.7. Actually, since the data is to one decimal place, the range is from 4.15 to 5.75: a width of 1.6. Four classes would have widths of <u>about</u> 1.6/4 = .4 each: *4.15-4.55, 4.56-4.95, 4.96-5.35, 5.36-5.75.* Eight classes would have widths of about 1.6/8 = .2 each: *4.15-4.35....* To have six classes the width would be 1.6/6 = .27 each: *4.15-4.42, 4.43-4.70,...* It is important to note that the division of the data into classes <u>is not</u> an exact science: there are no absolutes.

Walnut Creek Charlie Daniels

Wake	////// ////// ////
Orange	////// ////
Durham	////// /

If the data is not numerical, then it usually can be grouped by a certain attribute. For example, at a recent concert at Walnut Creek, 50 people were interviewed to determine the location of their residence. These people may be divided into classes designated by the county in which they live.

In both cases, once the classes have been defined, the frequency for each class is determined by counting the number of data units in each class.

Applying the Concept

a. Fifty people ran in a race. The times for the race need to be divided into 5 classes in order to make a histogram. The times (from lowest to highest) were *10.8, 10.9...12.7* What class intervals might be used to illustrate the data?

b. Forty workers were timed (in seconds) as they assembled a device. Their times (from lowest to highest) were *40, 41, 41, 42, 42, 42, ...52.* Decide on the number of classes you would use to construct a histogram and what the class intervals would be.

c. Ninety people from the high schools in Wake County attended an orientation session at Wake Technical Community College. What is a reasonable *nonquantitative* attribute that could be used to construct a histogram?

Summary

The steps to produce either of these graphics are similar.
1.   Gather and organize the data.
2.   Consider how many classes are to be used (usually 4 or 5 for the first 20 to 30 pieces of data and another class for every additional 20 to 30 pieces of data) and define the limits of each class.
3.   Determine the frequency for each class using some sort of tally.
4.   Produce the graphic.

Applying the Concept

Construct a histogram for the data shown. The data represents the top batting averages in the American League for 300 or more at bats during the 1995 season.

.354	.310	.305	.294	.278	.269	.256	.242
.332	.310	.302	.292	.278	.269	.256	.242
.332	.309	.301	.291	.277	.268	.255	.237
.323	.309	.300	.290	.277	.267	.254	.237
.323	.309	.300	.290	.277	.265	.253	.235
.322	.308	.299	.288	.277	.264	.252	.225
.320	.308	.299	.287	.277	.263	.250	.224
.319	.307	.298	.287	.275	.262	.250	.224
.316	.307	.298	.286	.274	.261	.250	.222
.314	.307	.297	.285	.272	.261	.248	.219
.314	.306	.296	.284	.271	.261	.247	.218
.314	.306	.295	.283	.270	.261	.244	.218
.313	.306	.295	.278	.270	.261	.244	.216
					.260	.243	.215

**The top 107 batting averages in the American League
at the end of the 1995 season.**

Construct a pie chart for the information shown.

How Your County Property Tax Dollar is Spent Wake County, 1995	Education	42.0%
	Health, Mental Health & Social Services	31.6%
	Sheriff, Jail, Emergency Medical Services and Emergency Mgmt.	8.9%
	General Administration	8.6%
	Capital Projects and Debt	4.1%
	Libraries	2.4%
	Parks and Recreation	0.4%
	Community Development	1.0%
	Insurance Programs	0.7%
	Other	0.3%

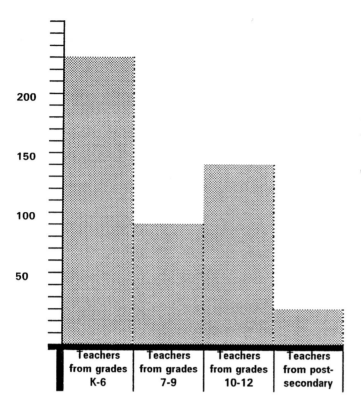

Take the information from the histogram to the left and construct a pie chart for the data. The data depicts the number of mathematics teachers from defined grade levels in attendance at a recent conference for mathematics educators.

When producing histograms and pie charts, be sure to carefully label the graphic. The true facts can easily be misrepresented by not labeling or mis-labeling axis.

_____ 3.2   Practice

1.   Construct a histogram for each of the following data sets.

Data set A: An experiment was performed which measured the number of hours a battery could run a toy rabbit. Twenty-four batteries were tested with the following results (in hours).

34 37 29 31 32 33 31 29 28 35 36 30
36 30 29 34 26 33 35 32 30 32 34 28

Data set B: The dosage in milliroentgens (mR) given by an X-ray machine is of critical importance. The dosage must be fairly constant and predictable. The dosage given by a machine was measured twenty times with the following readings:

4.25   4.26   4.26   4.21   4.44   4.43   4.33   4.50   4.25   4.28
4.30   4.31   3.90   4.44   4.41   4.35   4.20   4.27   4.41   4.40

Data set C: Find a fair dice and roll it 36 times. Record the result of each roll and make a histogram showing the frequency of each outcome.

Use a pair of dice and roll them 36 times. Record the results of each roll and make a histogram showing the frequency of each outcome.

Write a short paragraph explaining the differences in the two histograms and a possible explanation for those differences.

Data set D: The number of cellular telephone subscribers in the United States is rising quickly. Make a bar graph to show the number of subscribers for each of the past five years.

Year	1988	1989	1990	1991	1992
# (millions) of Subscribers	2.05	3.45	4.61	7.65	10.95

2.   Construct a pie chart for each group of data.

Data set E: Here are some facts about the proposed budget for the fiscal year, 1995. Make two pie charts, one for "Where the money would come from" and one for "Where the money would go".

Where the money would come from:
Excise taxes 5%
Corporate income taxes 9%
Social Security and other payroll taxes 32%
Other 15%
Individual income taxes 39%
Where the money would go:
Military 18%
Other Federal operations and grants to
    local and states 18%
Interest payment 14%
Social Security 22%
Retirement benefits 4%
Medicare 10%
Medicaid 6%
Unemployment benefits 2%
Other 6%

☺ Data Set F

The gross monthly salary for an accountant at C Count, Inc. is $3703.00. Construct a pie chart which shows the portion of the gross income for each category below as well as the "net take home."

Federal withholding $318.45
State withholding $130.36
Social Security $266.74
Retirement $222.18
Hospitalization $216.18
Car Payment $289.99

Data set G: The earth is surrounded by a sort of blanket of air. This blanket is composed of natural gases and traps the heat sent down from the sun, holds it around the surface of our planet, and keeps the air warm. These gases are known as "the greenhouse gases" because the blanket they form works like a greenhouse: it allows sunlight to pass through but holds in a certain amount of heat. The proportion of these gases has remained the same from the end of the last Ice Age until the beginning of industrial times. As a result, the earth's climate has remained relatively stable. Construct a pie chart showing the gases which are major contributors to the greenhouse effect: carbon dioxide (50%), chlorofluorocarbon and other halocarbons (20%), methane (16%), ozone (8%), and nitrous oxide (6%).

3.   Write a paragraph to tell why the advertisement at the beginning of this section may be misleading.

3.3   Experiments

Objectives of this section: The student will be able to
- *evaluate an experiment in terms of its intended outcomes.*
- *define error with regard to samples taken from a population.*

> Three sections of MAT 151 were being offered at 10:00 am. The instructors decided to do an experiment with the three classes. The instructor of section 01 decided to teach the class using all group work and allowing technology as needed, letting the students discover relationships and teach each other as the instructor coached each team. The instructor of section 02 decided to use lecture entirely with no technology. Finally, the instructor of section 03 taught the class as usual, with a careful blend of group work, lecture, and technology. At the end of two weeks an identical test was given each section. The class averages were 80, 77, and 82 for sections 01, 02, and 03, respectively.

The
Class
Experiment

Experiments

Experiments are very important to the researcher. Whether working in a pharmaceutical lab or on a construction lot, the outcome of experiments allows the worker to learn how to produce something better, quicker, or more efficiently.

> An *experiment* is the process by which measurements or observations are made.

3.31

Variability and Error

We shall use experiments to demonstrate the variability that exists when making random samples. Recall that the definition of random sample indicated that any sample was as likely to occur as any other. This indicates that the results are variable--that is, that the results are likely <u>not</u> to be the same for each sample. This logic would mean that if the Gallup Poll conducted two surveys on the same subject, they would be likely to get **different** results. This likely occurrence is why polls not only tell you the results but the "margin of error." The margin of error indicates the pollster is not sure about the result being absolutely the truth. The margin of error is associated with the variability of the population and the number in the sample taken.

> Any conclusion based on a sample of a population that is variable will have some degree of *error* based on the variability of the population and the number in the sample.

3.32

Applying the Concept

The results of a poll lead a pollster to say that he is 95% certain that Clyde Binton will get 48% (± *4% due to sampling error*) of the vote in next week's election. That means that the pollster is pretty sure that Binton will get between 44% and 52% of the vote. If the same data was used to predict *with 99% certainty* the percent of the vote that Binton would get, would the **margin of error** be more or less than 4%?

To learn more about experiments and variability, the class will perform several experiments.

Experiment #1

The class is given a container containing a significant number of beads of two colors, white and red (the total number of beads of each color is known by the instructor). The class will take 12 samples of 12 beads and record the number of white beads found in each sample. Record the results in the table to the right.
Based on the results, predict how many white beads are in the entire box.

Sampling Beads

Samples of 12	Number of White Beads
#1	
#2	
#3	
#4	
#5	
#6	
#7	
#8	
#9	
#10	
#11	
#12	
Average Number of White Beads Per Sample of 12	

Experiment #2

The class will be divided into two groups. Each group will try to unscramble a word in the shortest amount of time. The words will not be trivial and are about 6 to 9 letters in length. One group (the control group) is sent out of the classroom so that the other group (the experimental group) can receive the "treatment". The treatment is that the experimental group will learn that each word is related to the same subject. The other group is brought back into the room and the time it takes for each group to unscramble the word is recorded. What conclusion can you make concerning the treatment?

Experiment #3

The class will be told that it will determine the average horizontal distance a person can jump flat-footed by selecting and testing a sample of six jumpers from this class. Volunteers are asked to come to the front of the room and jump as far as they can. The distances are recorded and the average length is determined. Can the average found be advertised as the average distance people can jump?

Experiment #4

The class will take some scrap paper and wad it up so it can be thrown like a ball. Position four people about 8 to 10-feet from a waste basket located in the middle of the room. Give each person 8 "balls" and let them try to toss them into the waste basket. Repeat the process and record the average number of shots made on each trial.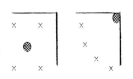

Repeat the experiment with the basket located in the corner of the room. If an experiment allows the conditions to change, what can be said about the results?

Experiment #5

A box without a top is used in this experiment. The bottom of the inside of the box is divided into six sections. Each section is designated as 0, 1, 2, 3, 4, or 5 points. A nickel is thrown into the box so that it will land randomly somewhere on the bottom of the box and within one of the designated regions. When the actual results differ from the theoretical results, there is usually a reason. What reasons can you identify for any discrepancy in this experiment?

On the histograms below, make a sketch of how you think the bottom of the box could have been divided into five regions <u>based</u> on the results shown in the histogram.

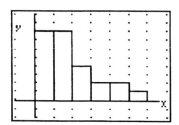

Select one of these experiments and write a paragraph about the experiment. Explain whether or not the experiment did what it was designed to do. Mention whether or not the experiment could be improved--and how.

3.4 Group Discussion and Work
Objectives of this section: The student will be able to
- *work in groups to perform an experiment.*
- *record the data for an experiment carefully and accurately.*

> The Secretary's Commission of Achieving the Necessary Skills (SCANS) listed several attributes that empowers expect from today's workers and which schools should help develop.

SCANS

1. The twenty-five members of a class listed below. The instructor has five problems to put on the board. <u>You</u> select the five people to go to the board using a random number generator. Repeat this process six times recording the number of females found in each group (their names are marked with asterisks).

Adam	Frank	Karl	*Pat	Ulysses
*Beth	George	Luke	*Queen	Victor
Carl	Hank	*Mamie	Robert	*Wanda
*Diane	*Ida	Ned	Sam	*Xylanda

Members of the Class

   a. Ten of the 25 members of the class are female. Do you think the class should suspect discrimination if the instructor calls on all of the five problems to be worked by females?

   b. Do you think the club members should suspect discrimination if none of the five members selected by the instructor are women? Explain.

2. Prior to the 1994 elections, a popular radio talk show host asked his callers to call in and vote for one of the Virginia Senatorial candidates: North or Robb. The response was enormous. Out of 1144 calls, 882 supported North. But in the actual election, North lost with only about 46% of the popular vote. What was wrong with the poll?

3. A sample survey asks a sample of 1324 adults whether they believe that life exists on other planets: 609 say yes. What percentage of the sample believe in extraterrestrial life? The polling organization announces a margin of error (holding with probability .95) of ± 3%. What conclusion can you draw about the percentage of all adults who believe that life exists on other planets?

☺4. Simulating a multiple choice test:
Working with another student, perform the following task. One student numbers his paper 1 to 20. That student then places one of the letters A, B, C, or D beside each number (as if the letter were an answer to a multiple choice question). When that is complete, the other students will call out one letter for each number. A right answer occurs when the guess matches the letter. Repeat the experiment three times. Construct a histogram that shows the frequency of correct answers on each of the four "tests."

The other student numbers his paper from 1 to 20 and places a letter beside each number as an answer to a question. The second student calls out <u>one</u> letter as the answer to every question. A right answer occurs when the guess matches the letter. Repeat the experiment three times. Construct a histogram similar to the previous one. Discuss the results.

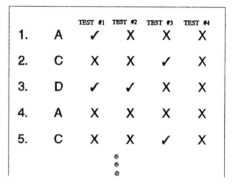

		TEST #1	TEST #2	TEST #3	TEST #4
1.	A	✓	X	X	X
2.	C	X	X	✓	X
3.	D	✓	✓	X	X
4.	A	X	X	X	X
5.	C	X	X	✓	X

5. A language teacher believes that the study of a foreign language improves the command of English. He examines the records at his high school and finds that students who elect to take a foreign language do indeed score higher on English achievement tests. What conclusions can he make?

6. A job-training program is being reviewed. Critics claim that because the unemployment rate in the manufacturing region affected by the program was 8% when the program began and 12% four years later, the program was ineffective. Are they correct?

7. The monthly salary of each employee in two companies is listed in the table below. You will take several samples from this population. Organize your work in a neat table to display your findings.

Take three random samples of 3 employees of each company and determine the average salary for each sample. Why are the results so different?

Monthly Salary of Employees in Two Companies (in $)				
Company A		**Company B**		
1500	1500	1400	1550	2400
1500	1500	1350	1800	2200
1500	1500	1550	1800	2100
1500	1500	1225	2250	2300
1500	1500	3000	1100	2400
1500	1500	4000	2050	2200
1500	1500	2500	3600	1950
		2400	2200	2100

Take three random samples of 6 employees of each company and determine the average for each sample. Are the results what you expected?

☺8. Research the SCANS report referenced at the beginning of this section. List three of the attributes the report identifies and that you feel are most important in order to be a successful employee.
[Via *Netscape* on the Internet: **http://www.jhu.edu:80/~ips/scans/** Secretary's Commission on Achieving Necessary Skills, (1991). *What work requires of schools: A SCANS report for America 2000.* Washington, DC: U.S. Department of Labor.]

3.5   Line Graphs

Objectives of this section: The student will be able to
- *construct a line graph from data.*
- *read and gather information from a line graph.*
- *construct a scatter plot.*
- *identify trends in a scatter plot.*
- *identify intervals in which a curve is increasing or decreasing.*
- *identify an optimum value from either a table or a graph.*

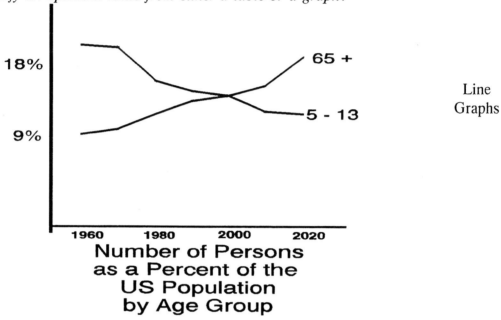

Line Graphs

**Number of Persons
as a Percent of the
US Population
by Age Group**

Line Graphs

Line graphs are often used to demonstrate the change in some variable as another changes. Colleges might use a line graph to demonstrate the increase in the beginning salary of their graduates over the last few years, or the increase in the cost of living over time compared to the small rise in instructor salaries over time. The type of graph usually used to demonstrate these relationships is a **line graph.**

> A *line graph* uses one or more lines to show changes in data. It shows how one or more variables change relative to another variable.

3.51

> A line graph is a *continuous curve* (one with no breaks) that connects the data points.

3.52

Horizontal and Vertical Axis

The horizontal axis or reference line in a line graph usually represents periods of time or specific times. The vertical axis or reference line is usually scaled to represent numerical amounts. Line graphs are used to present a picture of a trend in the data as well as maximum or minimum values.

The variables in line graphs may be something as different as dollars spent over time or the number of graduates for each year. But, in mathematics, these variables are related--one is an independent variable and the other a dependent variable. It is also customary to let the horizontal axis display the independent variable or the x-values. For that reason, the horizontal axis is called the x-axis. The vertical axis displays the dependent variable or y-values and is called the y-axis. Any point will have to have **two** values in order to determine a location--an x-value and a y-value. These two values are placed is alphabetical order and are called an *ordered pair*, (x, y).

Applying the Concept

Graph the following ordered pairs on the x-, y-axis shown to the right. Label each point with a letter.

A: (-3, 5)   B: (4, 2)
C: (0, 4)   D: (-2,-1)
E: (5, -2)   F: (-2, 0)

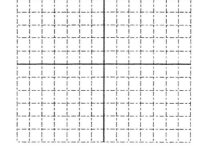

A line graph is constructed by connecting the ordered pairs from the smallest x-value to the largest x-value. Line graphs can show trends in the data. For example, a line graph can show the profit in a store month by month. In such a line graph, the stockholders would like to see the line graph increase as it moves from left to right.

Trends

A line graph shows the general trend as you move from the smallest x-value to the largest x-value. It is often beneficial to compare the *ratio* of the change is the y-values to the change in the x-values. This is done by selecting <u>two</u> points on the line graph, finding the change in y (*sometimes written as* Δy) from one point to another, finding the change in x (Δx) from one point to another and dividing Δy by Δx.

Consider the line graph shown here.

The vertical and horizontal lines connect the two points selected. There is a change in y of 11 and a change in x of 6 from one point to another. Therefore, the ratio of the change in y to the change is x is

$$\frac{\Delta y}{\Delta x} = \frac{11}{6}$$

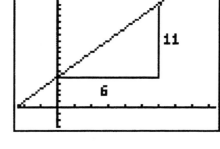

Change in y

Change in x

$$\frac{\Delta y}{\Delta x}$$

These changes usually have some kind of unit associated with them. The graph above could be the price of a share of stock from one day to the next. If that were the case, then the ratio would be $11 to 6 days.

Scatter Plot

If the data contains several y-values for any x-value or if the data is numerous, then a scatter plot is useful. The scatter plot shows all the data points in the set allowing the user to make conjectures about the data. The scatter plot can show where points are concentrated or provide a general trend in the data.

> The graph of all the ordered pairs in a set is a **scatter plot.**

3.53

Example

A class decided to take a survey of its members regarding the number of miles each person lived from campus and the number of minutes it usually took them to get from home to campus. The following data represents the information acquired in the survey. The data is represented in a scatter plot with the ordered pairs written as

(distance from campus in miles, average time of drive in minutes)

MILES	MIN.	MILES	MIN.
14	20	16	25
8	13	21	33
20	30	18	30
10	15	9	15
12	20	4	6
3	7	25	40
15	25	14	22
11	15	26	40
5	7	10	16
16	25	18	25

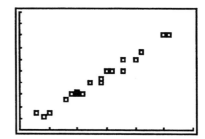

Scatter Plot (miles, minutes)
What statement can you make concerning the general trend you see?

Example

A local newspaper sponsors a five-mile run every fall. People of all ages participate-families, grandparents and kids. There are prizes for each age category. The scatter plot for a selected portion of the data is shown below. The ordered pairs represent **(age, time)**: that is, the age of the participant and the time it took for that person to run five miles. What statement can you make concerning the general trend in the data you see? Identify three of the ordered pairs. Were there more older or younger runners?

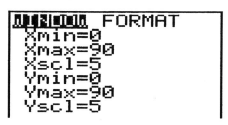

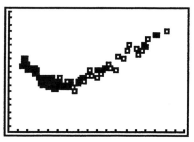

Scatter Plot - (age, time)

Trends

The following hypothetical data will be used to illustrate a line graph. The data represents the amount of time (in hours) it takes to paint a house (dependent variable) versus the number of people helping (independent variable).

Group Work

Construct a line graph for the data.

Data: (1, 11), (2, 6), (3, 4), (4, 3), (5, 2.5), (6, 2.25),
(7, 2.5), (8, 3), (9, 3.5), (10, 4), (11, 4.5), (12, 5)

Do you think that this line graph is an accurate representation of what would happen? Why?

> A curve is said to be *increasing over an interval* if everywhere within that interval, as you move from left to right, the curve moves upward. A curve is said to be *decreasing over an interval* if everywhere within that interval, as you move from left to right, the curve moves downward.

3.54

Consider another way of explaining *increasing*. Within the interval that a curve is increasing, select a point on the curve: (x_1 , y_1) . Now, move to the right and pick another x-value, still within the interval for which the curve is increasing: (x_2 , y_2). Obviously $x_2 > x_1$, but the point is that since $y_2 > y_1$ the curve is increasing.

So, if $x_2 > x_1$ and (x_1 , y_1) and (x_2 , y_2) are points on the curve, then $y_2 > y_1$ and the curve is increasing.

Write a sentence for a *decreasing* curve similar to the last sentence above for *increasing* curves.

Optimum Values

When a continuous curve changes from increasing to decreasing, it must achieve a maximum value. A minimum value is obtained when a continuous curve changes from decreasing to increasing. Being able to find these *optimum* values is an important task.

> An *optimum value* occurs when a continuous curve changes from increasing to decreasing (*a maximum value*) or when a continuous curve changes from decreasing to increasing (*a minimum value*).

3.55

Applying the Concept

Consider the table of values with the accompanying graph. Can you identify the approximate optimum value(s) numerically and on the graph?

Can you identify the interval over which the curve is increasing? decreasing?

Between what two x-values was the curve increasing the fastest? the slowest?

x-value	y-value	x-value	y-value	x-value	y-value
0.00	5.00	2.20	2.48	4.40	2.29
0.10	5.20	2.30	2.23	4.50	2.75
0.20	5.35	2.40	1.99	4.60	3.27
0.30	5.45	2.50	1.75	4.70	3.85
0.40	5.52	2.60	1.53	4.80	4.50
0.50	5.55	2.70	1.32	4.90	5.22
0.60	5.54	2.80	1.12	5.00	6.00
0.70	5.50	2.90	0.95	5.10	6.86
0.80	5.43	3.00	0.80	5.20	7.79
0.90	5.33	3.10	0.67	5.30	8.79
1.00	5.20	3.20	0.57	5.40	9.88
1.10	5.05	3.30	0.50	5.50	11.05
1.20	4.88	3.40	0.46	5.60	12.30
1.30	4.68	3.50	0.45	5.70	13.64
1.40	4.47	3.60	0.48	5.80	15.07
1.50	4.25	3.70	0.55	5.90	16.59
1.60	4.01	3.80	0.65	6.00	18.20
1.70	3.77	3.90	0.80	6.10	19.91
1.80	3.52	4.00	1.00	6.20	21.72
1.90	3.26	4.10	1.24		
2.00	3.00	4.20	1.54		
2.10	2.74	4.30	1.89		

```
RANGE
 xMin=0
 xMax=6.3
 xScl=1
 yMin=0
 yMax=16
 yScl=1
 y(x)= RANGE ZOOM TRACE G
```

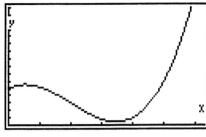

Find the Optimum Values

Today's problem solver needs to be able to approach a problem any one of three different ways: numerically (or with tables), graphically and symbolically. This is called the "three-pronged approach." You have examined this data using tables and a graph. The three-pronged approach to this problem will be completed by providing the symbolic relationship between the x- and y-values: $y = .4x^3 - 2.4x^2 + 2.2x + 5$. Which of these approaches provides you with the most information? Which is easiest to use? Explain.

A fourth approach is sometimes added: verbally. Can you describe in words what the picture of this graph looks like?

☺

Long-Term Project

Keep up with the odometer reading on your car for the next several weeks. Make a table with two columns: **DAY NUMBER** and **ODOMETER READING**. Mark today's date "DAY ZERO" and record your odometer reading beside zero. Every other day or so, record the odometer reading and number of the day. You will be asked to make a line graph of this data and use it for other problems later on. Keep good records!

Day Number	Odometer Reading

Ex 1 Many line graphs are done with time as the independent variable (along the x-axis). Construct line graphs for each data set below.

Data set A: Mickey Mantle played for the New York Yankee's baseball team for 18 consecutive years. Construct a line graph showing the number of home runs he hit in each of his 18 seasons.

Season	1	2	3	4	5	6	7	8	9	10	11	12	13	14	15	16	17	18
# of HR	13	23	21	27	37	52	34	42	31	40	54	30	15	35	19	23	22	18

Data set B: A strong cold front moved through the Raleigh/Durham area on November 10, 1994. The hourly temperatures reflect the change from an unseasonably warm morning to a cold evening. Construct a line graph for the hourly temperatures shown.

Time	6	7	8	9	10	11	12	1	2	3	4	5	6
Temperature	61	62	64	63	60	59	55	53	51	51	51	49	46

Data set C: The water tank for Knightdale is allowed to drain down to a certain level before the water is automatically turned on to refill the tank. This is usually done late at night when the water needs of other areas is a minimum. The water level (in feet from the bottom of the tank) is recorded each hour. Make a line graph showing the water level.

Time	6pm	7	8	9	10	11	12	1	2	3	4	5	6
Water level	22.1	21.9	20.5	20.3	20.0	19.9	19.9	19.8	20.2	20.8	21.4	22	22.2

Ex 2 A homeowner is concerned with the amount of energy used to heat the home. She keeps a record of the natural gas consumed over a period of nine months. Because the months are not all equally long, she divides each month's consumption by the number of days in the month to get the cubic feet of gas used per day. Then she goes to the weather service and records the number of degree-days for each month. (Degree-days are a measure of demand for heating: one degree-day is accumulated for each degree that a day's average temperature falls below 65 degrees F.) She divides this total by the number of days in the month giving the average number of degree-days during the month. Construct a scatter plot using the following data she collected. The x-axis will be the **Degree-days** in the month and the y-axis will be the **cubic feet** of gas consumed that month.

	Oct	Nov	Dec	Jan	Feb	Mar	Apr	May	June
Degree-days per day	15.6	26.8	37.8	36.4	35.5	18.6	15.3	7.9	0
Gas consumed per day (in cubic feet)	5.2	6.1	8.7	8.5	8.8	4.9	4.5	2.5	1.1

☺ Ex 3   A weight is placed on a spring and the distance the spring stretches from the position at which it rested with no weight attached is recorded. Make a line graph showing the change in length of the spring as weight is attached to it.

Weight (grams)	20	30	40	50	60	70	80	90	100
Distance (cm)	4	6	8	10	12	14	16	18	20

Ex 4   Plot the following points on a coordinate axis.
A: (3, 5)   B: (-2, -8)   C: (-5, 9)   D: (8, -2)   E: (0, 5)
F: (-2.7, -8.1)   G: (5.5, 0)   H: (0, 1.2)

Ex 5   The data listed in the chart below was collected from a thermometer placed inside a freezer. The time of zero represents the time of the initial temperature measurement. The data is in minutes and degrees Fahrenheit, respectively. How many minutes after the initial temperature measurement is minimum temperature achieved? What is the minimum temperature? After how many minutes is the maximum temperature achieved? What is it?

On which is the time of the maximum or minimum temperature more clear, the graph or the data? Why? Describe, in words, the shape of the curve?

Time	0	1	2	3	4	5	6	7	8	9	10
Temp.	18.7	20	18.5	14.7	10.0	5.9	4.1	4.9	8.3	12.9	17.3

Write an explanation of why the temperature varies and if you think this chart accurately reflects <u>the</u> maximum and minimum temperatures over the first 30 minutes.

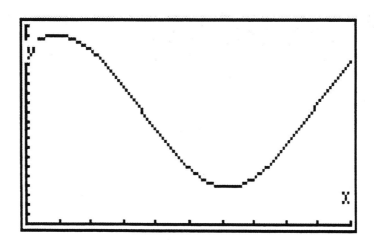

Ex 6   Plot and label the point (3,2) as point *P*. Now, moving only horizontally and vertically, move two units to the right and three units up. Label that point *Q*. From Q, move five units down and six units to the left; label that point *R*. From R, move three units down and two units to the left; label that point *S*. Write the coordinates of Q, R and S.

Ex 7   Plot and label the point (1,2) as point *E*. Now, moving only horizontally and vertically, move two units to the right and three units up. Label that point *F*. From F, move two units to the right and three units up; label that point *G*. From G, move four units to the right and six units up; label that point *H*. Write the coordinates of F, G and H. Why are all four points on the same LINE?

Ex 8   A machine at the local Coca-Cola plant fills bottles with 12 ounces of liquid. Of course, since even machines aren't perfect, it doesn't always put exactly 12 ounces in the bottle. Every 500th bottle that passes through the machine is taken out of production and the amount of liquid in the bottle is measured exactly. Graph #1 shows the result of these measurements on one particular day.

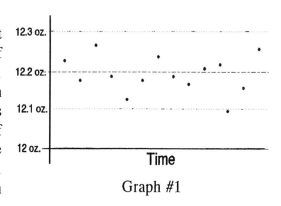

Graph #1

These scatter plots are used to maintain quality. A technician must make a decision regarding the production when the scatter plot identifies a problem. Hopefully, the technician can identify "trends" and make adjustments before production has to be stopped completely. Graph #2 shows a definite problem at the end of the day. What is the problem? When do you think the technician should have realized that something was amiss?

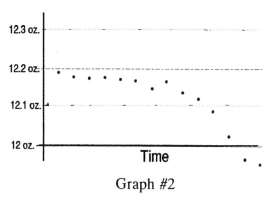

Graph #2

Ex 9   The following data might come from a paving company. While on a job, measurements were taken to record the depth of asphalt being laid. This particular job requires a 1.50 inch surface. During the job, there are several reasons why asphalt is not poured consistently. These measurements must be taken and recorded quickly in order to minimize the amount of rework needed. Plot the data, as was done is Graphs #1 and #2 above, and mark when you believe adjustments were made.

Sample	1	2	3	4	5	6	7	8	9	10
Depth (in.)	1.61	1.62	1.65	1.49	1.49	1.66	1.67	1.51	1.52	1.52

Ex 10 Pick one of the following graphs that most likely goes with the scenario presented. Write at least one sentence explaining your logic.

    a.   When the light turned green, Frances turned onto the Interstate and headed for the beach.

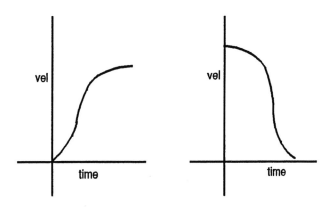

    b.   When David jumped off the platform, he felt weightless until the bungee-cord tightened just in time to pull him away from the oncoming ground.

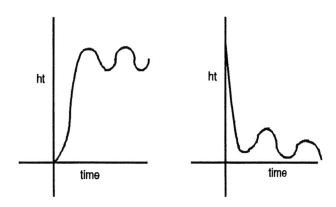

    c.   A car, with all its windows closed, was parked in direct sunlight. As the sun rose, the temperature inside the car began to climb.

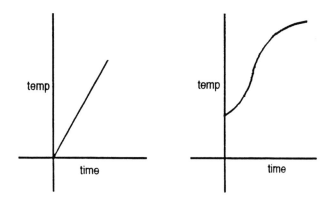

3.6 Using Technology

Objectives of the section: The student will be able to

- enter data onto a spreadsheet.
- use a spreadsheet to construct graphical displays of data.

As your college's facilities allow, your instructor will provide you with experience using a spreadsheet to construct line graphs, pie charts and histograms.

Select an appropriate way to represent the information in the following tables.

a. Types of Accidents
   i. Select one way to represent the data graphically.

b. The Presidential Election of 1912
   i. Represent the Popular Vote results with one graph.
   ii. Contrast the Popular Vote and the Electoral Vote using one graph.

c. The Median Age of the US Population
   Demonstrate how the median age of the US population has changed over the years from 1820.

d. Marriages and Divorces in the US
   Use a graphic that will allow the user to see the change in the number of marriages and divorces each year as well as to compare how they changed relative to one another.

☺e. The Presidential Elections of 1980 and 1984
   Construct a scatter plot of the ordered pairs showing the vote split to Democrats and Republicans in the southern states shown for each election. Make one graph for the 1980 election using Carter on the x-axis and Reagan on the y-axis. Make another graph for 1984. Use Mondale on the x-axis and Reagan on the y-axis. Are any states "out of line"?

f. The Electric Light company sampled 50 of their 60-watt bulbs in an effort to produce an advertisement concerning the average life expectancy of their best bulbs. Construct a histogram to show the results of the following data.

1200	1350	1410	1500	1300	1400	1310	1330	1425	1450
1350	1250	1510	1350	1400	1425	1475	1490	1440	1560
1390	1400	1415	1500	1480	1475	1410	1400	1500	1325
1290	1330	1400	1450	1525	1500	1375	1410	1450	1410
1310	1330	1400	1450	1500	1520	1550	1450	1440	1425

g. The Major Tire Company sampled 36 of their tires to find the average life span of the tire under normal wear and tear. It ran the tires on a moving belt in a factory and recorded the miles "driven" until the tread width was deemed unsafe. Construct a histogram for the data.

45000	47500	50000	46000	44000	52000
42000	51000	48000	45000	47500	48500
44000	45000	48000	46000	49000	42000
45000	48000	51500	40000	53000	49000
45000	48000	47500	50000	41000	47500

a) **Types of Accidents in the Home**
(National Safety Council)

Accident	No. of Occurrences
Falls	6100
Fires, burns	3900
Suffocation	1900
Poison (solid or liquid)	3100
Poison (gas)	800
Firearms	900

b) **Election of 1912**

Candidate	Popular	Electoral
(D) Woodrow Wilson	6,286,214	277
(PR) Theodore Roosevelt	4,216,020	88
(R) William Taft	3,483,922	8

c) **Median Age of the US Population**

Year	Median Age
1820	16.7
1840	17.8
1860	19.1
1880	20.9
1900	22.9
1920	25.3
1940	29.0
1950	30.2
1960	29.5
1970	28.0
1980	30.0
1982	30.6

d) Marriage and Divorce

Year	Marriages	Divorces
1984	2,487,000	1,155,000
1980	2,413,000	1,182,000
1970	2,158,802	708,000
1960	1,523,000	393,000
1950	1,667,331	385,144
1940	1,595,879	264,000
1920	1,274,476	170,505
1910	948,166	83,045
1900	709,000	55,751

e) Presidential Elections of 1980 and 1984
(Vote in Thousands)

State	1980 ELECTION		1984 ELECTION	
	(D) Carter	(R) Reagan	(D) Mondale	(R) Reagan
AL	637	654	552	873
FL	1419	2047	1448	2729
GA	891	654	707	1069
MS	430	441	352	582
NC	876	915	824	1346
SC	428	439	344	616

Chapter 4   Linear Functions

Many situations call for decisions to be made based on what is known about some variable or variables. Manufacturers make decisions concerning the number of items to produce based on many factors: demand and costs being two major factors. A mother buying clothes for her children decides to buy a certain number of items based on factors like the price of each item and her budget. And finally, the price you pay to fill up your car with gasoline is a function of the number of gallons you purchase and the price per gallon. For many relationships like this, there is a mathematical model to show how one variable changes as others do. In business and industry, the job of finding that model is often the responsibility of the mathematician. In this chapter, you, the mathematician, will learn to model relationships using a very important mathematical tool: functions.

4.1   Relations and Functions

Objectives of this section. The student will be able to

- *define and identify functions*
- *use functional notation*
- *find the domain and range of a function from ordered pairs, tables, or from graphs*
- *identify functions as discrete or continuous*

Going out of business sale! Business cars FOR SALE. Two Toyota Camrys: 1989 for $7000 and 1991 for $11000. Two Cadillac Sevilles: 1988 for $6000 and a 1993 for $16000. All four cars have been well cared for and have been driven about 20,000 miles each year.

Depreciation

Functions

The concept of function could easily be considered one of the most important ideas in mathematics. In order to understand functions, realize that relationships exist between the values of one variable and the resulting values of another variable. These two sets of data are somehow *related*. The relationship that connects a value in one set to a specific value in the other set is usually determined by some type of rule.

A *relation* is a correspondence between the elements in one set (the domain) and the elements in another (the range).

4.11

Below you will find elements of two sets. Draw lines between the sets to match the elements in the first set with a *related* element in the second set.

Domain: Albany   Ohio   :Range
                Cleveland   California
                Boston   New York
                Atlanta   Massachusetts
                Los Angeles   Georgia

A relation then simply conveys the fact that a relationship exists between two sets. An element from the first set may be "matched" with any number of the elements from the second set. You will quickly realize that this is not the ideal situation, especially in mathematics.

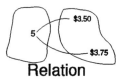

Relation

Consider the relationship that might exists between the number of gallons of gasoline purchased and the price you expect to pay. (For our purposes, let us say a gallon of gas cost $1.10.) The number of gallons purchased is related to the total price. If you purchase 10 gallons you would expect to hear "That will be $11.00 please." There is only one number corresponding with 10 that is appropriate. The idea that there is only <u>one</u> element in the range for any element in the domain is what determines that a relation is special--and is a function.

A *function* is a correspondence between elements in the domain and elements in the range such that exactly one element in the range corresponds to each element in the domain.

4.12

Determine if the relationships described are functional relationships.

a)   To every U S President, there corresponds a Vice President.
b)   To every number, there is a number less than that number.
c)   To every state, there is a capital.
d)   To every person, there is a birth date.
e)   To every number, there is a number that is always one more than the number.

f)

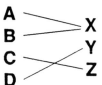

g)

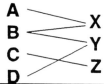

In mathematics we often have a *rule* that relates elements in the domain with elements in the range. That rule is often in the form of an algebraic expression. For example, if the element in domain is to be related to a number that is three more than twice the original number, then

$$y = 2x + 3$$

and 1 would correspond with 5, 2 with 7, 3 with 9 and so on.

Functional Notation

There is a special notation that we will use to denote functions.

> *Functional notation,* $y = f(x)$, will be used to denote that y is a function of x. This notation is read "y is a function of x."

4.13

This notation provides a way of denoting that y (the *dependent variable*) corresponds to some number x (the *independent variable*). Using this notation, f(1) is the value of the dependent variable corresponding to an x-value of 1. Using the notation on the example just above

$$f(x) = 2x + 3$$

Using this function, find f(1), f(2), and f(3).

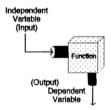

Independent Variable (Input)

Function

(Output) Dependent Variable

$f(x) = 2x + 3$	$f(x) = 2x + 3$	$f(x) = 2x + 3$
$f(1) = 2(1) + 3$	$f(2) = 2(2) + 3$	$f(3) = 2(3) + 3$
$f(1) = 2 + 3$	$f(2) = 4 + 3$	$f(3) = 6 + 3$
$f(1) = 5$	$f(2) = 7$	$f(3) = 9$

Using this function, the calculations above indicate that

1 is matched with 5
2 is matched with 7 and
3 is matched with 9.

> *NOTE: When substituting values for the variable, <u>use parenthesis</u> around the value you substituted for the variable.*

When functions involve fractions or decimals, remember to use the rules from arithmetic to evaluate the functions. Employ the rules of algebra when functions involve negative numbers and/or exponents.

Let C(n) = 1.35n + 2.50. Find C(3) and C(15).

$$C(n) = 1.35n + 2.50$$
$$C(3) = 1.35(3) + 2.50 = 4.05 + 2.50 = 6.55$$
$$C(15) = 1.35(15) + 2.50 = 20.25 + 2.50 = 22.75$$

Using the function C(n), 3 is matched with 6.55 and 15 is matched with 22.75.

Let $V(r) = (4/3)\pi r^3$. Find $V(6)$, $V(1.5)$

$$V(r) = \frac{4}{3} \pi r^3$$

$$V(6) = \frac{4}{3} \pi (6^3)$$

$$V(6) = \frac{4}{3} (216) \pi$$

$$V(6) = 288 \pi$$

$$V(6) = 904.78$$

Using this function,
$6 \rightarrow 904.78$
and
$1.5 \rightarrow 14.14$

$$V(r) = \frac{4}{3} \pi r^3$$

$$V(1.5) = \frac{4}{3} \pi (1.5^3)$$

$$V(1.5) = \frac{4}{3} (3.375) \pi$$

$$V(1.5) = 4.5 \pi$$

$$V(1.5) = 14.14$$

Group Work

a) Let $f(x) = 5x - 7$

   Find: $f(3)$, $f(-2)$, $f(.2)$, $f(0)$

b) Let $A(r) = \pi r^2$

   Find: $A(1)$, $A(3.5)$, $A(10)$

c) Let $V(t) = \frac{3}{4}t + \frac{1}{2}$

   Find: $V(10)$, $V(\frac{1}{4})$, $V(-1)$

Domain and Range

Functions, then, can be thought of as matching one number with another number--making ordered pairs. The elements that make up the set of first numbers is called the *domain*, and the elements that make up the set of last numbers is the *range*.

If a function matched the following elements together as shown in the ordered pairs, identify the domain and the range.

F: { (3,5), (5,9), (6,11), (9,17), (10,19) }

Domain: { 3, 5, 6, 9, 10 }
Range: { 5, 9, 11, 17, 19 }

Ordered pairs are readily graphed. So, if a function is shown as a set of *graphed* ordered pairs, you may still be able to determine the domain and range.

Each grid below is the "picture" of a function. That is, each grid contains the points that correspond to the ordered pairs that make up the function. The grids are scaled by 1 on each axis. For each function

   a)   list the ordered pairs for each function, and

   b)   identify the domain and range of each function.

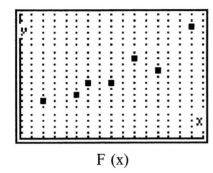

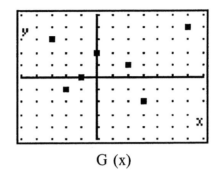

F (x)   G (x)

Notice that in these graphs, there is only *one* y-value for every x-value. This is the important characteristic of a function.

If a relation exists that is not a function, then there are two *different* ordered pairs with the same first element. Describe how that would appear on a graph.

Discrete and Continuous Functions

Functions may be classified as *discrete* or *continuous*. Discrete functions deal with data that is *countable*. Therefore, continuous data is not countable and can assume any value in one or more intervals on a line.

Functions that are discrete usually contain people or items. Functions that are measured in time are usually continuous.

A discrete function.
   An assembly line produces a device that requires 8 #12, 2-inch screws for each piece. This relationship can be written in functional notation: **N(s) = 8s**.
   Several ordered pairs from the function, written in the form **(s, N(s))**, are (1,8), (2,16), (3,24), . . . There would be no values for s = 1.2 or s = 2.8 -- just integral values are meaningful.

A continuous function.

> The length of a metal rod expands when heated. The original length is 10 inches at 70 degrees Fahrenheit. For every 25 degrees the temperature increases, the length increases by .02 inches. This relationship can be written in functional notation:
> **L(t) = .02(t- 70)/25 + 10.**
> Several ordered pairs from the function are (100, 10.024), (120.5, 10.04), (135.4, 10.052),... Notice that an ordered pair *may* exist for all values of t. [In reality, this model may only be appropriate for a small range of t.]

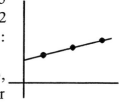

Applying the Concept

1. Example: A discrete function.

    When planning a party, a mother decides that she will need six balloons for each child coming to the party. Therefore, the number of balloons is a function of the number of children attending.

    a. Write a functional relationship that relates the dependent variable-number of balloons (*B*) to the independent variable-number of children (*C*).

    b. Find six ordered pairs that belong to this function.

    c. Is there a value for B(1.5)?

2. Example: A continuous function.

    In a recent weather segment, meteorologist Greg Fishel said that a cold front came through at 6:00 pm when the temperature was 66° F. The front caused the temperature to fall 12 degrees each hour for three hours.

    a. Write a functional relationship that relates the dependent variable-temperature (*F*) to the independent variable-time (*t*).

    b. Find six ordered pairs that belong to this function.

    c. Is there a value for F(1.5)?

---

1. Match the items in the first set with the items in the second set.

1.	Banana	Red
2.	Apple	Yellow
3.	Grapefruit	Green
4.	Strawberry	
5.	Butter Bean	

2. The following function is defined using a set of ordered pairs: either listed or on a graph. Identify the domain and range for each function.

a. {(3,5), (4,5), (5,7), (6,8)}

b. {(0,1), (1,0), (2,3), (2.1,3)}

c.

3. Evaluate each function using the indicated value of the variable.

Expression	x = 3	x = -2	x = .35	x = ¾
$f(x) = 3x - 5$				
$g(x) = x^2 - 1$				
$h(x) = \sqrt{x + .14}$				
$j(x) = \dfrac{x + 1}{x + 3}$				
$m(x) = 2x^2 + 3x - 5$				

4. Evaluate using functional notation.

   a. $v(t) = -16t^2 + 8t + 16$ *Find* $v(2)$

   b. $q(s) = 2(s + 3) + 3$ *Find* $q(-3)$

5. Let h(t) = 3t + 2. Find the range of h if the domain is {1,3,4,6}.

☺6. A recycling plant pays its customers for aluminum by the pound. One Saturday morning two customers each brought in one bag of aluminum cans. The first customer's bag weighed 24 pounds and he received $18.96.
   a. Find and write a mathematical model in functional notation for this situation.
   b. If the other customer's bag weighed 30 pounds, find the amount he should receive based on your model.

7. Using the information at the beginning of this section, how much are the two different cars depreciating each year? Can you write a model for each car to show the "value" in subsequent years?

_____   Reviewing The Basic Concepts

Some students may need to review some basic concepts of algebra.

The next two topics will be used in future work.

☺   Remember to continue to record your odometer readings every day or so. A sample journal would look like this.

Day Number	Odometer Reading
0	43455
3	43567
7	43688
12	43892

Algebra of Signed Numbers

Addition and Subtraction - The rules for adding signed numbers are demonstrated in the problems below. Notice that subtraction is changed to addition before performing any operation.

$$9+(-5) = 4 \qquad (-5)+3 = -2 \qquad (-6)+(-3) = -9 \qquad 5+(-8)+(-4) = 5+(-12) = -7$$

$$8-3 = 5 \qquad 3-9 = 3+(-9) = -6 \qquad (-5)-4 = (-5)+(-4) = -9 \qquad 2-(-6) = 2+(-(-6)) = 2+6 = 8$$

Try these. Do them step by step to be sure you do not miss any negative signs.

$$3 + (-6) =$$

$$(-5) + (-9) =$$

$$(-9) + (-7) + 4 =$$

$$(-9) - 3 =$$

$$(-3) - (-7) =$$

$$8 - (-3) - 8 =$$

Multiplication and Division - The rules for multiplication and division are demonstrated in the examples below. Division is changed to multiplication by multiplying by the reciprocal of the <u>divisor</u>.

$$9 * (-5) = -45 \qquad\qquad (-5) * (-3) = 15 \qquad\qquad (-6)(\tfrac{1}{3}) = -2$$

$$(8) \div (-4) = -2 \qquad\qquad (5)(-8)(-4) = (-40)(-4) = 160 \qquad (-30) \div (-6) = 5$$

$$18 \div \tfrac{1}{3} = \tfrac{18}{1} * \tfrac{3}{1} = 54 \qquad (-50) \div (\tfrac{2}{5}) = (-50)(\tfrac{5}{2}) = (\tfrac{-50}{1})(\tfrac{5}{2}) = -125$$

Try these:

$$80 \div (-4) =$$

$$(-40)(-3)(\tfrac{1}{6}) =$$

$$(-42) \div (\tfrac{1}{6}) =$$

$$(-\tfrac{4}{5}) * (\tfrac{20}{29}) =$$

$$(-20)\ (-3)\ (2) =$$

$$(-\tfrac{2}{3}) \div (-\tfrac{9}{4}) =$$

The order of operations demands that addition and subtraction be done before addition and subtraction. Try these:

$$\frac{8 - 2}{4 - 6} =$$

$$\frac{(-8) - (4)}{12 - (-8)} =$$

$$\frac{\tfrac{1}{2} - 3}{(-5) - \tfrac{3}{5}} =$$

Rearranging Equations

It is necessary to be able to rearrange the terms in an equation. The process of *isolating a variable* on one side of an equation is called *solving the equation for that variable.* The first step to solve an equation for a variable is to <u>isolate</u> the variable. That is done by moving all the terms that contain the variable to one side of the equation and all the other terms that do not contain the variable on the other side of the equation. After simplifying, the equation can be solved for the variable. Examples follow.

Solve for x:

$$3x + 5y = 15$$

$$3x + 5y - 5y = 15 - 5y$$

$$3x = 15 - 5y$$

$$\frac{3x}{3} = \frac{15}{3} - \frac{5y}{3}$$

$$x = 5 - \frac{5}{3}y$$

Solve for y:

$$7x - 3y = 9$$

$$7x - 3y + 3y = 9 + 3y$$

$$7x = 9 + 3y$$

$$7x - 9 = 3y + 9 - 9$$

$$7x - 9 = 3y$$

$$\frac{7x}{3} - \frac{9}{3} = \frac{3y}{3}$$

$$\frac{7}{3}x - 3 = y$$

Try these:

Solve for y: $10x - 5y = 25$

$9x + 8y = 16$

Solve for x: $7x + 21y = 3$

$.5x - 3y = 5$

4.2   Linear Functions-the Slope

Objectives of this section.   The student will be able to
- *estimate the slope of a line from its graph.*
- *find the slope of a line from two points or from the equation of the line.*
- *graph a line given the slope and any point.*

A roof is being built with a pitch of 5 to 12. That means that for every five feet you move horizontally, you move twelve feet vertically. If the distance from the edge of the roof to the center line is 15 feet, how tall will the roof be?

Pitch

Linear Functions

The graph of a *linear function*, as the name suggests, is a line. Recall that the graph of a function is made up of ordered pairs that make the function true. Therefore, all the ordered pairs that are produced by linear functions lie on the same line. Any subset of these ordered pairs, when written in tabular form, is called a *table of values*. In this section, we shall examine the attributes of graphs of linear function and the equations that produce them.

First, examine some points that are linear. Use a piece of graph paper to do the following examples accurately.

Example P)
Graph the point (-7,-9) and label it A. Move to the right 2 and up 3 and label this point B. Again move to the right 2 and up 3 and label this point C. Repeat the process one more time and label this point D. What makes these four points lie on the same line?

Example N)
Graph the point (6,3) and label is R. Move to the left 1 and up 2 and label this point S. Again, move to the left 1 and up 2 and label this point T. Repeat the process one more time and label this point U. Are all four points on the same line?

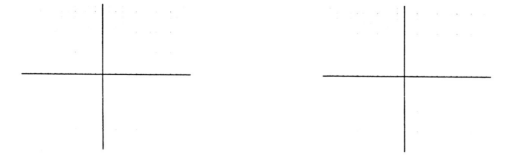

Slope

The point of the previous two examples is to demonstrate that a <u>constant</u> rate of change in both the x- and y-direction will produce a line. This rate of change in the y- and x-direction is an especially important concept. The ratio of the change in y divided by the change in x between any two points on any line is always a constant. This constant rate of change is called the *slope* of the line.

> The **slope** of a non-vertical line passing through the points (x_1, y_1) and (x_2, y_2) is
> $$slope = m = \frac{change\ in\ y}{change\ in\ x} = \frac{y_2 - y_1}{x_2 - x_1} = \frac{\Delta y}{\Delta x} \quad [x_1 \neq x_2]$$

4.21

Finding the Slope of a Line - Using Two Points

The slope of a line can be found if you know two points on the line. Simply use the formula shown above to determine the slope.

Warning: When calculating slope, always subtract the y values and the x values in the same order.

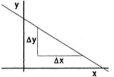

The change in the y, Δy, is often called the rise of the line between the two points. The change in the x, Δx, is often called the run. Using this terminology, the slope of a line is often heard being described as the *rise over the run*.

Slope is a number that describes the line's "steepness." The steeper the line the larger the slope. But, there are a couple of problems!

Group Work

Finding the slope of a line given two points:

Problem:

Lines that have the same degree of steepness but rise in different directions should not have the same slope! Notice that the lines in examples "P" and "N" rise in opposite directions. Lines which *rise* from left to right have a *positive slope* and lines that *fall* from left to right have a *negative slope*.

On a coordinate axis, pick two points so that when you draw a line through them the line rises from left to right. Use the slope formula to find the slope of the line.

Using the same axis, pick two points so that when you draw a line through them the line falls from left to right. Use the slope formula to find the slope of the line.

a) $(3, -4)\ (-2, 1)$

$m = \dfrac{-4 - (1)}{3 - (-2)}$

$= \dfrac{-5}{5} = -1$

b) $(-5, -7)\ (3, 9)$

$m = \dfrac{9 - (-7)}{3 - (-5)}$

$= \dfrac{16}{8} = 2$

Problem:

Lines that do not fall in either direction can not have a positive or negative slope! The slope of *horizontal* lines is zero. The slope of *vertical* lines is undefined.

On a coordinate axis, pick two points so that when you draw a line through them the line is horizontal. Use the slope formula to find the slope of the line. Is this a function? Can it be written in the form

$$f(x) = mx + b \text{ ?}$$

What are the value of m and b?

On a coordinate axis, pick two points so that when you draw a line through them the line is vertical. Use the slope formula to find the slope of the line. Is this a function? Can it be written in the form

$$f(x) = mx + b \text{ ?}$$

Explain.

A)   Pick any two of the four points you graphed in examples "P" and
      "N" and find the slope of the line through the four points.

B)   Find the slope of the line that passes through the points (4,-2) and
      (-2,6). Sketch the line. Does the sign of the slope match the
      direction in which the line rises?

C)   Find the slope of the line that passes through the points (0,-5) and
      (6,13). Sketch the line. Does the sign of the slope match the
      direction in which the line rises?

D)   If you calculate the slope wrong by subtracting the y and x values
      in a <u>different</u> order, can you describe the type of error you
      should *always* see?

Finding the Slope of a Line - Using The Equation of the Line
    The slope of a line can also be found from the equation of that line. One way to find
the slope of a line from the equation is to find two points on the line. For example,
consider the equation of a line

$$f(x) = y = 3x + 2$$

Since $f(1) = 5$ and $f(2) = 8$, two ordered pairs on the line are (1,5) and (2,8). The
slope of the line can be determined using the slope formula: m = 3. Since it does not
matter which two points you choose, choose x values that are easy!

$$m = \frac{\Delta y}{\Delta x}$$
$$= \frac{8 - 5}{2 - 1}$$
$$= \frac{3}{1} = 3$$

Examples

Find the slope of the line whose equation is given by finding two points and using the slope formula.

$$\text{a)} \quad 2x + 3y = 12$$

$$\text{b)} \quad 6x - 2y = 11$$

Consider $2x + 3y = 12$
Let $x = 0$, *then* $2(0) + 3y = 12$
$$3y = 12$$
$$y = 4$$
Let $x = 3$, *then* $2(3) + 3y = 12$
$$6 + 3y = 12$$
$$3y = 6$$
$$y = 2$$

Using the ordered pairs $(0,4)$ *and* $(3,2)$

The slope is $\quad m = \dfrac{2-4}{3-0} = \dfrac{-2}{3} = -\dfrac{2}{3}$ ■

Consider $6x - 2y = 11$
Let $x = 0$, *then* $6(0) - 2y = 11$
$$-2y = 11$$
$$y = \dfrac{11}{-2} = -\dfrac{11}{2}$$
Let $x = 1$, *then* $6(1) - 2y = 11$
$$6 - 2y = 11$$
$$-2y = 5$$
$$y = \dfrac{5}{-2} = -\dfrac{5}{2}$$

Using the ordered pairs $(0,-\dfrac{11}{2})$ *and* $(1,-\dfrac{5}{2})$

The slope is $\quad m = \dfrac{\frac{-11}{2} - \frac{-5}{2}}{0-1} = \dfrac{-\frac{6}{2}}{-1} = 3$ ■

Write a paragraph explaining how you would find the slope of a line from its equation using the method shown above. Include in your paragraph an explanation of why it does not matter which two x values you choose initially.

Your paragraph should have a topic sentence which is developed within the paragraph. Write complete sentences and use proper grammar.

There is another way to find the slope of a line given the equation. When an equation is written in the form $y = mx + b$, the coefficient of x, m is the slope of the line. Using the first example above, y = 3x + 2, notice that the slope indeed was 3, the coefficient of x.

Important! The coefficient of x is the slope of a line *if and only if* the equation is that of a line <u>and</u> the equation is solved for y in terms of x.

Find the slope of the two lines shown in the previous two examples by solving the equation for y in terms of x.

For $2x + 3y = 12$
$$-2x \qquad = -2x$$
$$3y = -2x + 12$$
$$\tfrac{1}{3}(3y) = \tfrac{1}{3}(-2x + 12)$$
$$y = \tfrac{-2}{3}x + 4$$

Therefore, the slope is $\tfrac{-2}{3}$.

For $6x - 2y = 11$
$$+ 2y = + 2y$$
$$6x \qquad = 2y + 11$$
$$-11 = \qquad - 11$$
$$6x - 11 = 2y$$
$$\tfrac{1}{2}(6x - 11) = \tfrac{1}{2}(2y)$$
$$3x - \tfrac{11}{2} = y$$

Therefore, the slope is 3.

1. Graph the two points (*2, -5*) and (*-4, 7*). After graphing the two points but before calculating the slope make a conjecture as to whether the slope will be positive or negative. Find the slope of the line through these two points. Which of the following equations is probably the equation for this line?

   y = 2x - 9   y = -2x - 1   y = ½x - 6   y = -½x - 4

2. The equation of a line is *y = 4*. Find two ordered pairs that make this equation true. Use the slope formula to find the slope of the line. What is the coefficient of x in this equation? Does this agree with the slope you calculated? Graph this line. Describe the line's direction.

3. The equation of a line is *x = 2y + 3*.
   a)   Find the slope from the equation.
   b)   Find two ordered pairs on this line and that make this equation true. Use the slope formula to find the slope of the line using these two points.
   Do the two values you found as the slope agree?

4.   The equation of a line is $x = 2$. Find two ordered pairs that make this equation true. Use the slope formula to find the slope of the line. Can you find the slope from this equation? Explain. Does this agree with the slope you calculated? Graph this line. Describe the line's direction.

Summary

The graph of a linear functions is a line.

Functions written in the form

$$f(x) = y = m\,x + b$$

are linear functions; the coefficient of x, m is the slope of the line. Slope is defined as the rise over the run as the line moves from one point to another. Therefore, if you know one point on the line and the slope of the line, other points on the line may be found by "moving" from the given point "over and up" to the next point so that the "rise over the run" is the slope of the line.

Functions written in this form are symbolic models that represent relationships between two variables.

The slope can be determined from the equation or by using the slope formula and two points on the line.

$$slope = m = \frac{change\ in\ y}{change\ in\ x} = \frac{y_2 - y_1}{x_2 - x_1} = \frac{\Delta\,y}{\Delta\,x} \quad [x_1 \neq x_2]$$

Horizontal lines have a slope of zero.

The equation of a horizontal line looks like

$$f(x) = y = (0)\,x + b$$
$$\text{or}$$
$$y = b.$$

Vertical lines have a slope which is undefined; they are not functions and cannot be written in $f(x)$ form. The equation of a vertical line looks like the following:

$$x = c$$

1. Find the slope of the line passing through the given two points.
   a)   (3, 2) and (-2, 3)   d)   (2, 4) and (-3, 4)
   b)   (0, 4) and (-2, -2)   e)   (2, 1) and (2, -2)
   c)   (½, 1) and (¼, -3)

2. Find a second point on a line that
   a)   passes through (1, 2) with a slope of 2.
   b)   passes through (-1, 3) with a slope of 3.
   c)   passes through (-2, -3) with a slope of -4.
   d)   passes through (1, 3) and is horizontal.
   e)   passes through (-2, 3) and is vertical.

3. a)   Graph a line with a slope of 4 passing through the point (4,5).
   b)   Graph a line with a slope of -3 passing through the point (5,2).

☺4. Make a table of values and graph the line whose equation is
   $y = \frac{3}{5} x + 4$

☺5. Graph the line whose equation is x = 2.

☺6. Graph the line whose equation is y = 3.

7. Find the slope of the line whose equation is
   a)   y = 5x - 3   c)   x = 9
   b)   8x + 4y = 1   d)   x = 3y - 2

8. Without graphing the points shown in the table of values, can you tell if these points represent a line? How?
   a.

x	y
-2	1
0	3
1	5
3	7
7	9

   b.

x	y
-2	-4
0	2
1	5
3	11
7	23

☺9. The slope formula can be written in the form you see to the right. The Greek letter Delta (Δ) means "change in" so the slope can be viewed as the change in y over the change in x. This concept of "change in" is very important--it connotes a *rate*. The idea that slope is the *rate of change in one variable as another changes* is useful.

$$Slope$$

$$m = \frac{\Delta y}{\Delta x}$$

Describe the *rate* found in each problem below. You must use units to accurately describe a rate!

a) A company uses straight-line depreciation for a copy machine. The initial cost was $6000 and will be depreciated over 5 years to a scrap value of $1000.

b) A family on vacation traveling down I-95 is averaging 65 mph. They are currently 250 miles from their destination.

c) A typist typed a 1200 word paper in 30 minutes.

10. The roof (at the beginning of the section) extends 15 feet on the front and 20 feet on the back. What will the "pitch" of the back roof be?

4.3   Linear Functions-Intercepts and Applications

Objectives of this section. The student will be able to
- *find the x- and y-intercept from the equation of the line.*
- *identify the intercepts from the graph of a line.*
- *understand and apply the slope as a rate of change of one variable with respect to another in applied problems.*

> When John went to college, his Dad gave him $500 to use as he needed. Several weeks into the quarter, John looked back at his daily expenses and found that he had averaged spending $17 per day.

Send
Money!

Intercepts

There are two obvious points on any line with a slope other than zero --the intercepts. The point at which a line crosses the y-axis is called the *y-intercept*, and the point at which the line crosses the x-axis is called the *x-intercept*. The x-value of any ordered pair on the y-intercept is *zero,* and likewise the y-value of any ordered pair on the x-axis is *zero*. These facts provide a way to find each intercept.

> The *y-intercept* of a line is the point (0,b) where the line intersects the y-axis. To find b, substitute 0 for x in the equation of the line and solve for y.
>
> The *x-intercept* of a line is the point (a,0) where the line intersects the x-axis. To find a, substitute 0 for y in the equation of the line and solve for x.

4.31

Examples:

1.   Find the x- and y-intercepts of the line whose equation is
$3x + 4y = 12$.

$3x + 4y = 12$	$3x + 4y = 12$
To find the x-intercept, let y = 0:	*To find the y-intercept, let x = 0:*
$3x + 4(0) = 12$	$3(0) + 4y = 12$
$3x = 12$	$4y = 12$
$x = 4$	$y = 3$
$(4,0)$	$(0,3)$

2. Find the x- and y-intercepts of the line: $f(x) = 1.5x + 3.3$.

$f(x) = 1.5x + 3.3$

$f(x) = 1.5x + 3.3$

To find the x-intercept,
let y = 0:

To find the y-intercept,
let x = 0:

$(0) = 1.5x + 3.3$
$-3.3 = 1.5x$
$-2.2 = x$

$(-2.2, 0)$

$f(x) = 1.5(0) + 3.3$
$f(x) = 0 + 3.3$
$f(x) = 3.3$

$(0, 3.3)$

<u>IMPORTANT</u>
The coefficient of
x is the slope
<u>only</u> when the
equation has been
solved for y.

This last example provides us with the "rest of the story." That is, when an equation is written in the form $f(x) = y = mx + b$, the coefficient of x, *m* is the slope and the constant term; *b*, is the *y-intercept*. That is why the

$$y = mx + b$$

form of an equation of a line is called the ***slope-intercept*** form of the equation of a line!

When an equation is written in the form $y = mx + b$, the graph of that equation is a line. The slope of that line is the coefficient of x, ***m***, and the y-intercept of that line is the constant term, ***b***.

4.32

Group Work

1. Write the equation of the line in the slope-intercept form and give the slope and the y-intercept of the line.
    a. $2y - 3 = x$
    b. $4x + 2y = 3$

2. Find the x- and y-intercepts of the line whose equation is given.
    a. $f(x) = 2x + 3$
    b. $2y - 5x = 10$

The intercepts are usually good points to plot in order to graph a line. They are easy to solve for and relatively easy to locate. They also can be used rather easily to find the slope of the line since the axes provide the rise and run in an obvious fashion. The figure to the right shows the rise and run after the two intercepts have been plotted.

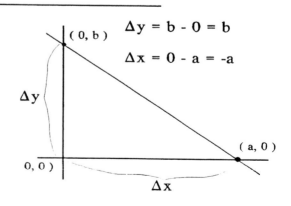

Using the Intercepts to find the Slope

Graph the line by finding the x- and y-intercept.
Find the slope of each line.

1. $2x + 3y = 12$

2. $5x - 2y = 10$

3. $f(x) = 2x - 1$

Applications of the Slope

The slope of a line has many applications as a rate of change. The intercepts also have application. In the following applications the independent and dependent variables are not necessarily x and y. The term x-intercept may still be used as long as it is understood to mean the intercept on the horizontal or independent axis. The y-intercept is where the line crosses the vertical axis. The y-intercept is f(0) in functional notation.

Depreciation

A computer purchased for $4,000 is expected to depreciate according to the formula $V(t) = -800t + 4000$. Where the V(t) is the value of the computer after t years. When will the computer be worthless?

The y-intercept in this application is the beginning value of the computer $4,000. The computer is worthless when its value, V(t) is zero. This is the x-intercept. The slope in this application represents the rate of depreciation: 800 dollars per year.

$$V(t) = -800t + 4000$$
$$0 = -800t + 4000$$
$$800t = 4000$$
$$t = 5 \text{ years}$$

Loans

A person owes $7,200 (including interest). The debt will be paid off by paying $900 per year. The debt balance can be calculated according to the formula $D(t) = 900t - 7200$. When will the loan be paid up and the person out of debt?

The y-intercept in this application is the beginning debt, $7,200. The debt is paid off when D(t) = 0, or at t = 8, the x-intercept. The slope in this application represents the rate at which the debt is paid, or 900 dollars per year.

$$D(t) = 900t - 7200$$
$$0 = 900t - 7200$$
$$-900t = -7200$$
$$t = 8 \text{ years}$$

Fuel consumption

The fuel compartment on a generator holds 3.6 gallons of gasoline. The generator burns fuel at the rate of .8 gallons per hour. The amount of fuel left in the tank can be shown by the formula

$$g(t) = -.8t + 3.6.$$

The y-intercept in this application is the initial amount of fuel: 3.6 gallons. The fuel will be used up when g(t) = 0, or at 4.5 hours; the x-intercept. The slope in this application represents the rate the fuel is used or .8 gallons per hour.

$$g(t) = -.8t + 3.6$$
$$0 = -.8t + 3.6$$
$$.8t = 3.6$$
$$t = 4.5 \; hours$$

Profit

The profit for a musical is dependent on the number of tickets sold at $12 per ticket and the total expenses. The total expenses for the musical were $2424. The profit can be found according to the equation P(n) = 12n - 2424. How many tickets will it take to break even?

The y-intercept in this application is the profit (loss) if no tickets were sold: -$2424. The break-even point occurs when the expenses minus the revenue equals zero; the x-intercept. The slope in this application is the rate that the revenue comes in or 12 dollars per ticket.

$$P(n) = 12n - 2424$$
$$0 = 12n - 2424$$
$$12n = 2424$$
$$p = 202 \; people$$

Group Work

Graph each line in each of the four examples above. Label and scale the axes appropriately. Clearly identify the x- and y-intercepts.

1. Find the x- and y-intercept for each line whose equation is given. Use the two intercepts to find the slope of the line.
    a.   2x - 3y = 18
    b.   5y - 15 = 3x

2. Find the slope and the y-intercept for each line whose equation is given in functional notation.
    a.   f(x) = 3x - 7   c.   p(r) = .08r - 1
    b.   g(n) = 1.2n - 9.1   d.   c(t) = 3

☺3. Find the x- and y-intercept and use them to sketch the graph of the line whose equation is given.
    a.   4x - 3y = 12   b.   .2x + .3y = 6

☺4. In each of the following applications, find the slope and intercepts. Define each using units as in the examples in this section. That is, write a paragraph and put in words what the slope and intercepts mean.

    a.   Hourly Wages.
    A worker's hourly earnings, *E(h)* can be calculated according to the formula: E(h) = 4.5h.

    b.   Descent
    The height in feet of a parachute, *H(t)* can be found according to the formula: H(t) = 2500 - 40t

    c.   Interest
    The monthly balance in an account drawing interest, *B(m)* can be found according to the formula: B(m) = 10m + 2000

    d.   Demand
    The number of VCR's consumer buy, *D(p)* depending on the price of the VCR, can be found according to the formula: D(p) = -.75p + 175

    e.   Supply
    The number of VCR's that a manufacturer will produce, *S(p)* depends on the price according to the formula: S(p) = -.75p + 100

    f.   Gears
    The rotational speed of a large gear *(V)* with *24* teeth is related to the speed of a smaller gear *(v)* with 6 teeth, *V(v)* by the following formula:

    $$V(v) = \frac{6v}{24}$$

4.4 Linear Functions-Finding the Equation

Objectives of this section. The student will be able to
- *find the equation of a line given two points*
- *find the equation of a line given a point and the slope*
- *use a linear model to extrapolate and interpolate*

A mathematical model can be described as an equation which models a physical phenomenon. Provided there is enough information about a linear function, the mathematical model for that function should be found in the form of a line.

Equations of lines can take several forms. The general form of an equation of a line is the least useful: $ax + by = c$. The slope-intercept form of an equation of a line provides the slope and the y-intercept: $f(x) = y = mx + b$. A third equation for a line is called the point-slope form: $y - y_1 = m(x - x_1)$.

General Equation	$ax + by = c$	Equations
Point-Slope Form	$y - y_1 = m(x - x_1)$	of
Slope-Intercept Form	$f(x) = y = mx + b$	Lines

Equations of Lines

In order to determine a unique line, two parameters must be provided. Either you know two points on the line, or you know a point and the slope of the line. In either case, the equation of that line can be found.

> *Remember: If an ordered pair is <u>on</u> the line, then when the x- and y-values from the ordered pair are substituted in the equation of the line, the equation is <u>true</u>.*

Example

Find the equation of the line through the point (4,2) with a slope of 3.

a) Using the slope-intercept form of the equation of a line:

$$f(x) = mx + b$$
Since the slope is 3, m = 3.
Since (4, 2) is on the line,
when x = 4 and y = 2
the equation is true.
We know everything except b:
$$2 = 3(4) + b$$
$$b = -10$$
Therefore: $f(x) = 3x - 10$

b. Using the point-slope form of the equation of a line.

$$y - y_1 = m(x - x_1)$$
Using the x- and y-value from the ordered as well as the slope:
$$y - 2 = 3(x - 4)$$
$$y - 2 = 3x - 12$$
$$3x - y = 10$$

While the two equations above do not look exactly the same, they still model the same line: the one through (4,2) with a slope of 3. It will also be necessary to be able to write an equation of a line in another form.

Three more examples:

    a. Find the equation of the line that passes through the two points (2, -2) and (5, -4).

Find the slope of the line:
$$m = \frac{-2 - (-4)}{2 - 5}$$
$$m = \frac{2}{-3} = -\frac{2}{3}$$
Use the slope and one of the points in the slope-point form of the equation of the line.
$$y - (-2) = -\frac{2}{3}(x - 2)$$
$$3y + 6 = -2x + 4$$
$$2x + 3y = -2$$

    b. Find the equation of the line that passes through the two points (1,5) and (1,7).

Since the two points lie on the same vertical line, the equation of the line is in the form x = a and the equation of the line is
$$x = 1$$

    c.   Find the equation of the line through (0,4) with a slope of -5.
The point (0,4) is the y-intercept. Since we know the slope, m = -5, the appropriate form to use is the slope-intercept form.
$$f(x) = -5x + 4$$

Group Work

1. Change the equation written in the general form to the slope-intercept form of an equation of a line.
    a. $3x - 8y = 24$
    b. $.2x + .25y = 1$

2. Change the equation written in the point-slope form to the slope-intercept form of an equation of a line.
    a. $y - 3 = 5(x - 2)$
    b. $y + 2.25 = .25(x - 4)$

☺ 1. Find the equation of the line using the given information. Graph the line.
  - a. A line that passes through (3,2) with a slope of 4.
  - b. A line that passes through (1,-2) with a slope of -2.
  - c. A line that passes through the points (95,35) and (41, 5).
  - d. A line that passes through the points (16, 4) and (25,1).
  - e. A line that passes through the points (4,2) and (4,8).
  - f. A line that passes through the points (0,2) and (2,0)

☺2. Straight-line depreciation is a type of depreciation in which the amount of depreciation taken each year remains constant. In terms of a linear function, the depreciation can be thought of as the slope of a line that relates the time, t to the remaining value, v. For example, the initial cost of a computer was $6,000. If the computer is to be depreciated using the straight-line method over 5 years and the salvage value is determined to be $500, then the yearly depreciation is $1,100. The equation for the remaining value is

$$value\ (time) = v(t) = 6000 - 1100t$$

Determine the function $v(t)$ [value(time)] for each example below.

A) Initial cost $5000
    Number of years   10
    Salvage value $500

B) Initial cost $70,000
    Number of years   30
    Salvage value $10,000

☺ *Remember to keep recording your odometer reading.*

4.5   More on Linear Functions-Applications and Models

Objectives of this section. The student will be able to
- *find a linear model from a problem stated in words*
- *use an algebraic model to construct answers for a verbal problem.*

Use your skills learned about finding equations of lines to find a linear function to model the following phenomenon.

Step 1: Read the information carefully.
Step 2: Determine the dependent and independent variables.
Step 3: Define, in writing, the variables in the problem.
Step 4: Identify the two parameters provided in the information.
Step 5: Determine the equation for the line.

☺1.   A well drilling company charges its customers by the number of feet it digs into the ground. Use the information below to find a linear model for the cost function.

A well that was 350 feet deep cost $2200.00.
A well that was 200 feet deep cost $1300.00.
How much would be charged for a well that was 250 ft deep? 400 feet deep?

Steps to # 1:

1: Read the problem.

2:    The dependent variable is cost and the independent variable is number of feet.

3: Let f = number of feet and c = cost

4: (350, 2200)
   (200, 1300)

2.   While in Boston for a conference, four mathematics teachers took a cab from the conference center to dinner. On the way back, the group had to take another route in order to see several of the more popular sites in Boston. If the trip to dinner was 4.8 miles and cost $10.40 and the return trip was 6.2 miles and cost $12.85, find a model for the taxi cab cost. How much would a 10.0 mile trip cost?

3.   You are probably familiar with how a clamp works. You thread a screw through a hole which advances the screw and thus closes the clamp. The distance the clamp closes is a linear function that is dependent upon the "pitch" of the screw. If a clamp advances 2 mm for every complete revolution (the pitch is 2 mm), find a model which will provide the amount of opening in the clamp as it is closed if it started with a 200 mm opening.

4.   The velocity of a box sliding down a ramp is a function of time. Use the following two measurements to determine a model that will show the velocity of the box at any time *t*. After 1.00 seconds, the velocity was 12.2 ft/s. After 4.50 seconds, the velocity was 35.4 ft/s. How fast do you think the box will be traveling after 5.00 seconds?

☺5. The velocity of an object falling from a high building is measured at two different times. After 1.22 seconds, the object was traveling at 23.6 ft/sec. After 1.30 seconds, the object was traveling at 26.78 ft/sec. How fast do you think the object was traveling after 1.25 seconds? How fast do you think that the object was traveling after 2.5 seconds? If it is discovered that the object was traveling at 100 ft/sec after 2.5 seconds, how you would explain the discrepancy?

6. From 1960 until 1980, when a large company moved into a small town, the population of that town grew constantly at a rate of 50 people per year. If the population of the town in 1970 was 2000, find a model for the population of the town for the years between 1960 and 1980.

In 1984 the population was 4000, in 1988 it was 4600, and in 1992 it was 5600. Discuss the trend in the population after 1980.

7. A linear model is to be used to model some data. But when three of the data points are collected, they don't lie on the same line! How many different lines could model this data if you selected two points at a time? How many lines could model 4 points, no three of which were on the same line? 5 points? 6 points? 20 points?

8. The three points first considered in Ex 7 were A:(6.2, 4.8), B:(7.2, 5.3), and C:(7.4, 5.7). If the linear model is derived using only points A and B, how far from the line is point C?

4.6 Other Functions

Objectives of this section. The student will be able to
- construct a table of values for a non-linear function.
- graph a function given the equation.
- compare and contrast two models.

Example 1
The height of a projectile can be approximated over time using the following function.

$$Height\ (time) = H(t) = V_0 t - 16t^2$$

V_0 is the initial velocity of the projectile.

Complete the table of values for
$$V_0 = 40\ ft/sec.$$

Use appropriate x- and y-scales to plot these points on a graph.

Write a short paragraph about your graph, particularly pointing out what you think is happening over the last two or three points and what you think will happen as you increase time.

| Projectile with an initial velocity of 40 feet/second ||
Time (seconds)	Height (feet)
0	
.25	
.5	
.75	
1	
1.25	
1.5	

Example 2
The monthly payment, P, for a 30-year mortgage depends on two factors: the amount of the loan, A and the monthly interest rate, r.

$$Monthly\ Pmt(Int\ Rate) = P(r) = \frac{Ar(1 + r)^n}{(1 + r)^n - 1}$$

For a 30-year loan, $n = 360$.

Therefore, formula below is for a 30-year loan of $100,000 is

$$P(r) = \frac{100000r(1 + r)^{360}}{(1 + r)^{360} - 1}$$

Complete the table shown by finding the monthly payment for a loan of $100,000 using the formula above for each interest rate shown.

Plot the points from the table and determine if the points are linear. What did you learn about the effect of the interest rate on the monthly payment of a mortgage?

| Monthly payment for a Mortgage of $100,000 |||
Annual Interest Rate	Monthly Interest Rate (r)	Monthly Payment (P)
6 %	.005000	
6½ %	.005417	
7 %	.005833	
7½ %	.006250	
8 %	.006667	
9 %	.007500	
10%	.008333	
12%	.01	

☺ Example 3

Imagine that you have a rectangular piece of cardboard that measures 18 inches by 24 inches. You can form an open box by cutting congruent squares from each of the four corners of the cardboard and folding up the flaps. The volume of the resulting open box is a function of the size of the square cut from each corner. (Recall that the volume of a box is found by multiplying the length times the width times the height.)

If you cut a square that measures x by x from each corner, then the dimensions of the open box will be x by $18 - 2x$ by $24 - 2x$. The volume of the box is given by the function

Volume of the Box (size of the square cut) =
$$V(x) = x(18 - 2x)(24 - 2x)$$

Volume of a box cut from an 18-inch by 24-inch piece of cardboard	
Size of the square (x)	Volume of the Box
1	
2	
3	
4	
5	
6	
7	

Complete the table of values shown using the formula for the volume of the box. Use the values to make a sketch of the function

Write a paragraph to discuss the volume of the box as the value of x changes. Discuss the possibility of a maximum volume possible as well as what will happen as the value of x continues to increase.

Does each square have to be the same size, Why?

Example 4

Mr. Brown works for U Sell It, Inc. His salary is based on the amount of sales he produces, on his *commission*. During any month, the percentage of commission he receives is based on the amount of sales; the higher the sales, the higher the commission.

The·commission is 3% if the total sales are $0 - $5,000.

The commission is 3.5% if the total sales are $5,001 - $10,000.

The commission is 4% if the total sales are $10,001 - $25,000.

The commission is 4.25% if the total sales are over $25,000.

Amount of monthly commission based on sales	
Amount of Monthly Sales	Commission
$2000	
$4000	
$6000	
$8000	
$10,000	
$15,000	
$22,000	
$27,000	
$32,000	

Complete the table to the right using the information above. Make a sketch of the function by connecting the points.

Write a paragraph about the shape of this graph and comment on why the graph "bends?" Is that good or bad for the salesman?

1. Repeat example #1 at the beginning of this section but use an initial velocity of $V_o = 50$ ft/sec and then $V_o = 75$ ft/sec. Plot the two sets of points on a graph and compare the result using a short paragraph.

Projectile with an initial velocity of 50 feet/second	
Time (seconds)	Height (feet)
0	
.25	
.5	
.75	
1	
1.25	
1.5	

Projectile with an initial velocity of 75 feet/second	
Time (seconds)	Height (feet)
0	
.25	
.5	
.75	
1	
1.25	
1.5	

2. Repeat Example #2 using a constant rate of 9% but allowing the amount of the loan to vary. Plot the results.

Monthly Payment for a Loan at 9%	
Loan Amount	Monthly Payment (P)
$120 000	
$110 000	
$100 000	
$90 000	
$80 000	
$70 000	
$60 000	
$50 000	

Volume of a box cut from an 20-inch by 30-inch piece of cardboard	
Size of the square (x)	Volume of the Box
1	
2	
3	
4	
5	
6	
7	
8	
9	

3. Repeat Example #3 but let the rectangular piece of cardboard be 20 inches by 30 inches.

4.7   Electronic Spreadsheets-taking the pain out of functions

Objectives of this section. The student will be able to
- *construct a table of values and a line graph using a spreadsheet.*

Hands-on class time will be provided for students to construct simple tables for functions.

☺For the portfolio problem, construct a line graph to go along with each table.

☺ Spreadsheet	
x	.06x + 100
175	
200	
225	
250	
275	
300	
325	
350	
375	
400	
425	

Spreadsheet	
x	.35 x + 2
5	
10	
15	
20	
25	
30	
35	
40	
45	
50	
55	

☺ Spreadsheet	
x	200 - 2x
5	
10	
15	
20	
25	
30	
35	
40	
45	
50	
55	

☺ Spreadsheet	
x	$3x^2 + 2x - 5$
1	
2	
3	
4	
5	
6	
7	
8	
9	
10	
11	

Spreadsheet	
x	$\dfrac{2x + 3}{3x - 5}$
0	
1	
2	
3	
4	
5	
6	
7	
8	
9	
10	

Spreadsheet	
x	$.066x - .02x^3$
0	
.1	
.2	
.3	
.4	
.5	
.6	
.7	
.8	
.9	
1	

Chapter 5   More Statistics-Linear Regression

Today's society demands that you understand more about data and the terms that are used to describe it. Numerical data is gathered in the workplace, in the market, in politics, and in sports. This chapter will examine the basic ways that statistics is summarized and examine a significant application.

5.1   Measures of Central Tendency

Objectives of this section. The student will be able to

- *understand and compute measures of central tendency*
- *appropriately use measures of central tendency to describe data sets*

A report in a local paper indicated that the median selling price for homes in Wake County rose 3% from the previous month to $112,500.

Median
Price

Analyzing Data

Numerical data can be acquired in many different ways. Large amounts of data must be organized before being analyzed. The first step is usually to list the data in ascending or descending order so that a minimum and maximum value can be determined easily. This process is known as *sorting the data*. Sorted data is more efficient to use than reading through long unorganized lists of numerical data.

The *range of the data* is determined by subtracting the minimum value from the maximum value--the highest minus the lowest.

Measures of Central Tendency

Three other statistical descriptions about data can be classified as *measures of central tendency*. They, in different ways, describe the tendency of the data to be grouped around the middle. The three measures of central tendency are the *mean, median and mode.*

Mean

The *mean* is defined as the arithmetic average.

> ### Mean
>
> $$mean = \frac{sum \ of \ the \ data \ entries}{number \ of \ data \ entries}$$
>
> *or symbolically as*
>
> $$\bar{x} = \frac{x_1 + x_2 + x_3 + \ldots + x_n}{n}$$
>
> *where x_1, x_2, x_3, ..., x_n represent the data entries and n represents the number of data entries.*

5.11

Median

The *median* is the value that falls in the center of the sorted data set.

> *The median is the value which falls in the middle when the data is arranged in ascending or descending order. The median is found by first sorting the data.*

5.12

○   If there is an odd number of data entries, the value which falls in the middle is the median.

○   If there is an even number of data entries, the average of the two middle values is the median.

Mode

The number which occurs with the greatest frequency is the mode.

> *The mode is the data value that occurs most often.*

5.13

If there are no values which occur more than once, there is no mode for the data set. Some data sets may have more than one mode.

A. One Monday night, Wake Medical Center had ten babies born in 15 minutes. The babies' weight (to the nearest tenth of a pound) were

5.5, 6.2, 10.0, 6.7, 4.2, 5.8, 6.9, 7.3, 8.2, 8.2.

The data, sorted from smallest to largest, is

4.2, 5.5, 5.8, 6.2, 6.7, 6.9, 7.3, 8.2, 8.2, 10.0

The range is 10.0 - 4.2 = 5.8 lbs.

The mean is

(4.2+5.5+5.8+6.2+6.7+6.9+7.3+8.2+8.2+10.0)/10 = 6.9 lbs

The median is (6.7 + 6.9)/2 = 6.8 lbs

The mode is 8.2 lbs.

B. The weights of the next 7 babies (already sorted) were

4.6, 5.2, 5.3, 5.9, 6.2, 7.0, 8.4

The mean is

(4.6+5.2+5.3+5.9+6.2+7.0+8.4)/7 = 6.1 lbs.

The median is 5.9 pounds.

There is no mode for the data.

_____ Applying the Concept

Speeding

A policeman was stationed on a stretch of road on which the residents had complained about cars speeding through the area. The posted speed limit is 45 mph. The data below represents the speed (in mph) of vehicles passing the policeman during one hour of the day.

Speeds: 44, 54, 50, 46, 45, 49, 55, 44, 42, 55, 51, 52, 54, 44, 60, 44,
59, 41, 44, 47, 51, 56, 44, 44, 47, 42, 43, 44, 59, 45, 52, 53

Sort the data and find the maximum values, minimum values and the range.

Find the mean, median and mode for the data. If you lived in that area and wanted to promote more enforcement of the speed limit, which statistic would you choose to argue your case? Which would you not choose? Why?

Is the road very busy? Explain your reasoning.

Applications

As you can see, different choices of measures of central tendency are more useful in certain applications. A real estate broker may talk more about the median price of homes in the area than the mean price. A manufacturer may talk more about the mean production level since there are probably not any outliers--or extreme values. A clerk in an apparel shop would be more interested in the mode of dress sizes sold to predict future orders.

_____ Extending the Concept

The mean can be described with concrete objects. Consider the following numbers as a data set:

$$1, 4, 4, 7, 9, 11$$

We can think of each of the numbers as being a weight placed on a balance. Each number should have the same weight. Where would the fulcrum go to balance the weights? Obviously, six, since six is the mean.

A fulcrum is the support for a beam or lever.

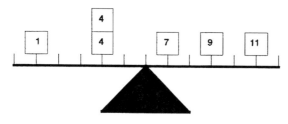

Now, consider this set:

$$5, 6, 6, 6, 7, 12$$

The mean is seven; the fulcrum is at seven. Aside from that, the picture of the data set is different. How would you describe the differences in the two pictures that represent these two data sets?

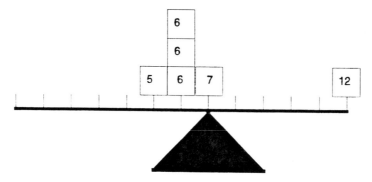

Graphical Means

The following set of drawings display a method of finding the mean from a histogram. To find the mean of the frequencies in a histogram, simply find the height at which all the class frequencies would be the same if you could rearrange the bars and move area from the tallest to the shortest bars.

These three illustrations show that the mean of 7, 8, 10, 12, and 13 is 10.

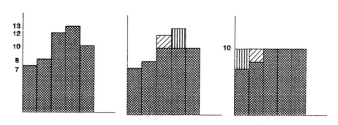

The Average is 4

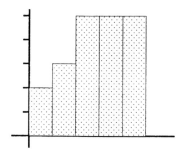

Original Data:
2,3,5,5,5

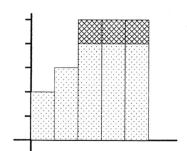

Select the rectangles
to be moved

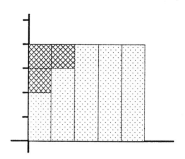

Move the rectangles to form
one large rectangle

The Average is 7

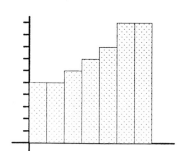

Original Data:
5,5,6,7,8,10,10

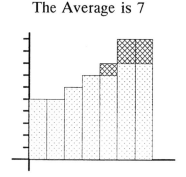

Select the rectangles
to be moved

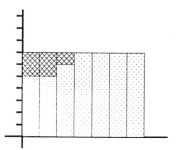

Move the rectangles to form
one large rectangle.

1. In an experiment on the effect of a drug on reaction time, a subject is asked to depress a button whenever a light flashes. Her reaction time for 12 trials are (in milli-seconds)
    96, 101, 102, 138, 93, 99, 107, 93, 95, 100, 101, 101
    Find the mean, median, mode, and range for this data.

2. In 1798, the English scientist Henry Cavendish measured the density of the earth in a careful experiment with a torsion balance. Here are his 23 repeated measurements of the same quantity (the density of the earth relative to that of water) made with the same instrument.
    5.36, 5.57, 5.44, 5.27, 5.63, 5.75, 5.29, 5.53, 5.34, 5.39, 5.34, 5.68, 5.58, 5.62, 5.79, 5.42, 5.46, 5.85, 5.65, 5.29, 5.10, 5.47, 5.30
    Find the mean, median, mode, and range for this data.

☺3. The number of consecutive hours a light bulb will burn before it burns out is tested. The number of hours for 39 bulbs is shown below. Find the mean, median, mode, and range for this data.
    400, 402, 403, 404, 410, 422, 422, 424, 425, 428, 429, 430,
    430, 431, 440, 442, 444, 444, 446, 446, 448, 450, 452, 455,
    456, 459, 460, 462, 463, 466, 469, 472, 476, 477, 480, 484,
    485, 490, 505

4. A small company has six employees, including the owner. Find the mean, median, and mode for the six salaries in this company.
    $25,050, $26,750, $32,000, $33,500, $42,000, $95,250

☺5. Considering the vignette at the beginning of the chapter, why is the "median price" used when papers talk about home prices in Wake County?

☺6. Perform an experiment with three other people in your class.
Roll one dice 12 times and record the number of times a six is rolled. Each person is to perform this experiment 10 times and record the number of times a six is rolled. Each person is to find the mean, median, and mode for their data. Then, all 40 numbers from all four people are to be put together. Find the mean, median, and mode for that data.

7. Find the mean of the following numbers: 5, 6, 18, 23, 24, 25, 25

Now, take a regular 36-inch yardstick and place any coin (all the same) on the locations that correspond to the numbers in this list. A paper clip can be used to secure the coins to the yardstick. That is, center a coin on the five-inch mark, one on the six-inch mark, etc, until you have all seven coins in place. Hold the yardstick carefully with one outstretched finger, so that it is supported on the 18-inch mark. Is the yardstick balanced?

Where would you place one other coin in order to keep the yardstick balanced by holding it on the 18-inch mark if two other coins were placed on the yardstick, one on the nine-inch and one on the ten-inch marks?

Now remove all the coins and place four coins on the yardstick: one on the five-inch, one on the six-inch, one on the 20-inch, and one on the 29-inch mark. Where do you have to hold the yardstick in order to keep it balanced? Think carefully about the results!

Discuss your findings in paragraph form.

8. Construct a histogram for the data below. Then, "move" enough rectangles to show that the average is 8.

Data: 6   6   7   8   9   12

5.2   Standard Deviation

You have been assigned the task of buying several 60-watt light bulbs. The average life of a 60-watt light bulb produced by the two different companies you may deal with is the same--the price is also the same. If you were a buyer for a company that used 1000 of these bulbs each month, what other information would you want to know before making your choice?

Watt?

Objectives of this section. The student will be able to
- *understand and compute measures of variation*
- *appropriately use measures of variation to describe data sets*
- *use a spreadsheet to calculate the standard deviation*

Measures of Variability

The measures of central tendency indicate, to some degree, the position of the data. They provide indications about the center of the data. As we compare data sets, we must also be able to indicate the degree of variability within the data set. The range is one way to indicate the degree to which a data set is *spread out*. The larger the range is, the larger the spread of the data. However, this leaves something out! Is the larger range a result of one or two outliers or is most of the data spread out? In order to communicate the variability of the entire set, we must look at each piece of data, not just the two extremes. The *standard deviation* of a data set measures variability associated with each of the data entries. For this reason, it requires many calculations and is not easy to compute by hand if the data set is large.

The symbol Σ means "*sum of*" and is often found in mathematics.

$$\sum_{i=1}^{4} i = 1+2+3+4 = 10$$

$$\sum_{i=1}^{3} 2i = 2+4+6 = 12$$

$$\sum_{i=1}^{3} i^2 = 1+4+9 = 14$$

*The standard deviation is a measure of variability.
The <u>variance</u> is the sum of the squares of the
differences between each data entry and the
mean divided by the number of data entries.
The <u>standard deviation</u> is the square
root of the variance.*

5.21

$$Variance = \frac{sum\ of\ (x_i - \bar{x})^2}{number\ in\ the\ data\ set} = \frac{\sum(x_i - \bar{x})^2}{n}$$

$$Standard\ Deviation = \sqrt{variation} = \sqrt{\frac{\sum(x_i - \bar{x})^2}{n}}$$

Standard Deviation

Here are the steps to follow to compute the standard deviation.

1. Compute the mean of the data set.
2. Compute the difference between each data value and the mean of the data set.
3. Square each of the differences.
4. Find the sum of the squared differences and divide the sum by the number of data entries. This number is the variance of the data set.
5. Compute the square root of the variance.

When doing the number of calculations required to find the standard deviation, it is helpful to organize your work in a table like the one shown below.

These calculations may be done on many calculators and, most assuredly, on a spreadsheet. But in order to fully understand this concept and what the number that represents the standard deviation means, our calculations will first be done by hand. It is convenient to perform this task in a tabular format.

Example

Find the standard deviation of the weights of ten babies used in 5.1:

5.5, 6.2, 10.0, 6.9, 4.2, 5.8, 6.9, 7.2, 8.1, 8.2.

In this example the mean is 6.9 pounds. The table to the right is used to find the differences between each data entry and then the squared differences.

The variance is the sum of the squared differences divided by n:

$$\frac{24.78}{10} = 2.478$$

Therefore, the Standard Deviation is

$$\sqrt{2.478} = 1.57 = 1.6 \ lbs$$

Weights (Data Entries)	Differences	Squared Differences
4.2	4.2 - 6.9 = -2.7	$(-2.7)^2$ = 7.29
5.5	5.5 - 6.9 = -1.4	$(-1.4)^2$ = 2.96
5.8	5.8 - 6.9 = -1.1	$(-1.1)^2$ = 1.21
6.2	6.2 - 6.9 = -.7	$(-.7)^2$ = .49
6.9	6.9 - 6.9 = 0	$(0)^2$ = 0
6.9	6.9 - 6.9 = 0	$(0)^2$ = 0
7.2	7.2 - 6.9 = .3	$(.3)^2$ .09
8.1	8.1 - 6.9 = 1.2	$(1.2)^2$ = 1.44
8.2	8.2 - 6.9 = 1.3	$(1.3)^2$ = 1.69
10.0	10.0 - 6.9 = 3.1	$(3.1)^2$ = 9.61
Mean = 6.9	Sum of the Differences = 0	Sum of the Squared Differences = 24.78

The standard deviation indicates how spread out the data is. The larger the standard deviation, the more spread out the data is. To get some understanding of this, let's see how many of the data entries are *within one standard deviation of the mean.* One standard deviation above the mean (6.9 lbs.) would be the value

The mathematical shorthand for adding and subtracting the same two numbers is sometimes used in cases where you must find one standard deviation *above* and *below* the mean.

6.9 lbs. + 1.6 lbs. = *8.5 lbs.*

One standard deviation below the mean (6.9 lbs.) would be the value

6.9 lbs - 1.6 lbs. = *5.3 lbs.*

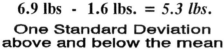

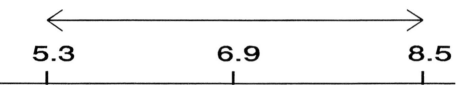

Place an "x" under the line and in the appropriate location for each piece of data within these bounds.

Use the figure below to indicate how many of the data entries are *within two standard deviations of the mean.* Place an "x" below the line for each piece of data.

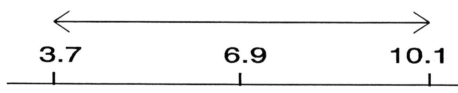

Normal Data

For "normal data" (a phrase we will define better later), about two-thirds or 67% of the data should be within one standard deviation of the mean. Most, or approximately 98%, of the data will be within two standard deviations of the mean.

Normally, when large samples are taken from data, the data is concentrated around the mean with fewer data found as you get further from the mean.

Applying the Concept

Consider a very large sample of "normally distributed" data; take the heights of 1000 30-year old men. If this data was displayed on a histogram over the range of 60 to 80 inches, what do you think the general shape of the histogram would be like? Can you sketch the curve that would connect the very tops of the bars on the histogram?

Consider another example.

Example

In this example the mean is 6.0 pounds. The table to the right is used to find the differences between each data entry and then the squared differences.

The variance is the sum of the squared differences divided by n:

$$\frac{21.72}{10} = 2.172$$

Therefore, the Standard Deviation is

$$\sqrt{2.172} = 1.47 = 1.5 \; lbs$$

Find the numbers associated with one standard deviation above and below the mean. What percentage of data entries occur within one standard deviation of the mean?

Find the numbers associated with two standard deviations above and below the mean. What percentage of data entries occur within two standard deviations of the mean?

Weights (Data Entries)	Differences	Squared Differences
4.2	4.2 - 6.0 = -1.8	$(-1.8)^2$ = 3.24
4.5	4.5 - 6.0 = -1.5	$(-1.5)^2$ = 2.25
4.8	4.8 - 6.0 = -1.2	$(-1.2)^2$ = 1.44
5.2	5.2 - 6.0 = -.8	$(-.8)^2$ = .64
5.9	5.9 - 6.0 = -.1	$(-.1)^2$ =.01
6.1	6.1 - 6.0 = .1	$(.1)^2$ = .01
6.2	6.2 - 6.0 = .2	$(.2)^2$ .04
6.7	6.7 - 6.0 = .7	$(.7)^2$ = .49
6.8	6.8 - 6.0 = .8	$(.8)^2$ = .64
9.6	9.6 - 6.0 = 3.6	$(3.6)^2$ = 12.96
Mean = 6.0	Sum of the Differences = 0	Sum of the Squared Differences = 21.72

Find the standard deviation of the following sets of numbers.

    a) 35, 37, 40, 42, 46

    b) 25, 35, 40, 45, 55

    c) 10, 39, 40, 41, 70

    d) 32, 42, 42, 42, 42

A spreadsheet is an obvious tool for finding the standard deviation of data sets. Although many spreadsheets have a "built-in" function for standard deviation, the spreadsheet will be written to perform calculations similar to the ones done by hand when finding the standard deviation. We can check this result against the spreadsheet's built-in formula. The spreadsheet below indicates the formulas and information that is placed in each cell.

This spreadsheet shows the formulas that might be used to build a template to find the standard deviation. The formulas necessary for different software packages might be different.

Data Set: 5, 8, 8, 9, 10

	A	B	C
1	$^\wedge x$	$^\wedge x$ - mean	$^\wedge (x - mean)^\wedge 2$
2	5	+a2-a7	(b2)^2
3	8	+a3-a7	(b3)^2
4	8	+a4-a7	(b4)^2
5	9	+a5-a7	(b5)^2
6	10	+a6-a7	(b6)^2
7	@sum(a2...a6)/5	@sum(b2...b6)	@sum(c2...c6)
8	'Variance		+c7/5
9	'Standard Deviation		@sqrt(c8)

☺ 1. Follow the five steps to find the standard deviation of each example.

a) 9, 10, 14, 15, 18, 19, 20

b) 15, 17, 18, 18, 19, 19, 19, 19, 20, 21, 21, 22

c) 8.23, 9.14, 9.23, 10.41

2. Consider the following data set:

8, 10, 12, 14, 16

The mean is 12.

(a) By changing the entry 16 to 21, the mean is changed to 13. What <u>one</u> change can you make to make the mean 10?

(b) The standard deviation of the original data set is 2.82. By changing the 16 entry to 12, the standard deviation is reduced to 2.04. Can you change <u>one</u> entry to make the standard deviation greater than 3?

(c) Can you change one entry to make the standard deviation less than 2?

5.3 Box-and-Whisker Plots

Doctors directed nurses to administer a hematology test every day to a patient undergoing treatment. The data represents the lymphocyte (LYMPHO) and white blood cell (WBC) count each day.

Day	1	2	3	4	5	6	7	8
LYMPHO	5.1	5.2	5.1	5.0	5.3	3.5	5.2	5.5
WBC	14	15	13	12	16	15	14	15

If you examined this data, would there be any questions?

Serious
Business

Objectives of this section. The student will be able to
- *find the percentiles and quartiles of a data set*
- *construct a box-and-whisker plot*
- *evaluate the z-score for a piece of data.*

Analyzing Observations

Data sets are usually measurements taken on a sample or a population. Whether the measurements are taken by humans or read from a complicated device, common sense must be used when inconsistencies occur in the data. An unusual observation that lies outside the range of the data values is called an *outlier*.

Outliers may be attributable to one of several causes. First, the outlier may simply be recorded incorrectly. Second, the outlier may be the result of a misclassified measurement. That is, the measurement was not taken from the population being studied. Finally, the measurement may be invalid. For example, a measuring device may have malfunctioned, the data may have been mixed with other data or the information coded incorrectly in the computer. These data are very large or very small relative to the other data. Since they tend to skew a histogram and they usually require further study.

An observation or measurement that is unusually large or small relative to the other values in a data set is called an **outlier.**

5.31

Box-and-whisker plot

One procedure for detecting outliers is to construct a *box-and-whisker plot (or boxplot for short)* for the data. Boxplots are constructed using *quartiles*. Quartiles are used to show the relative position of a particular measurement in a data set. A measure that expresses the position of the measurement in terms of a percentage is called a *percentile* for the data set.

Let x_1, x_2, x_3 ..., x_n be a set of measurements arranged in <u>increasing</u> order. The *pth percentile* is an element of the data set, x_q, where x_q separates the data into two groups: data at or below x_q and the data above x_q; such that **at least** *p%* of the measurements fall at or below the *pth* percentile.

5.32

The median, by definition, is the 50th percentile. The 25th percentile, the median, and the 75% percentile are often used to describe a data set. These divisions are called *quartiles*, the *lower quartile (25%)*, the *median (50%)*, and the *upper quartile (75%)*. They are represented symbolically by Q_L (the lower-quartile), Q_M or M (the mid-quartile), Q_U (the upper-quartile).

Finding Quartiles and Percentiles for Small Data Sets

1. Rank the n measurements in increasing order of magnitude.
2. Calculate the quantity $\frac{1}{4}(n + 1)$ and round **down** to the nearest integer. The measurement with this rank represents the lower quartile or 25th percentile.
3. Calculate the quantity $\frac{3}{4}(n + 1)$ and round **down** to the nearest integer. The measurement with this rank represents the upper quartile or 75th percentile.

To find the *pth* percentile, calculate the quantity *p(n + 1)/100* and round **down** to the nearest integer. The measurement with this rank is the *pth percentile.*

5.33

Example

24 High School Graduates

Suppose our data consisted of salaries of twenty-four high school graduates entering the workforce directly after graduation.

```
12400  12600  13000  13000  13500  13600
14000  14200  14400  15000  15500  16000
16500  18000  18500  18500  19000  19500
19500  22000  22500  25000  25500  29000
```

Since 25% of (24 + 1) rounds down to 6, then the sixth entry represents the 25% percentile: 13600. Note that 6 entries fall at or below 13600, and 18 entries (75% of 24) fall above 13600.

Since 75% of (24 + 1) rounds down to 18, then the eighteenth entry represents the 75% percentile: 19500. 19 entries fall at or below 19500.

A *box-and-whisker plot* is a figure that uses the quartiles to show the relative position of the data. The boxplot is made up of two components: the whiskers and the box. The *whiskers* extend to the lowest and highest data entry. The *box* extends from the lower-quartile to the upper-quartile with a vertical line at the median.

Examples

Here are two examples of boxplots. The first is a boxplot of the data in the example: **24 High School Graduates**. The second is a boxplot from the example: **Speeding** (earlier in the chapter).

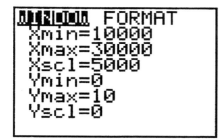

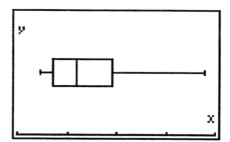

24 High School Graduates

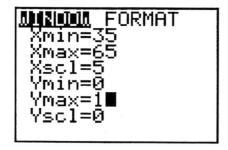

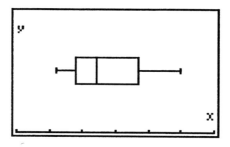

Speeding

There are 32 pieces of data in **Speeding**. The lower-quartile is the eighth piece: 44. The median is the 16th piece: 47. The upper-quartile is the 24th piece: 53.

Note

[There are different "rules" for establishing the quartiles. We will be consistent and use the rules written here. Boxplots are designed to demonstrate the variability of data. For small samples, those in which the number of data is small enough to allow the actual data to be examined easily, boxplots are not needed. The rules for establishing quartiles in this section may produce a percentage of the data significantly different from the percent called for if the sample is small. But, since we do not need boxplots for small samples, that really is not a concern.]

Z-scores

Another measure of *relative standing* is the *z-score*. A z-score identifies the position of a piece of data to the mean relative to the standard deviation. Those values less than the mean have a <u>negative z-score</u> and the values above the mean have a <u>positive z-score</u>. The z-score is the *number of standard deviations* a value is from the mean.

It can be computed using the formula:

$$z = \frac{x - \bar{x}}{s}$$

Where s is the standard deviation of the sample, $\bar{x}$ is the mean and x is the value under scrutiny.

Example

1. The mean for the **24 Graduates** data is 17529. The standard deviation for this sample is 4560.
   a. The value 14400 is below the mean and will have a negative z-score: $z = (14400 - 17529)/4560 = -.69$
   b. The value 22500 is above the mean and will have a positive z-score: $z = (22500 - 17529)/4560 = \mathbf{1.09}$

2. The mean for the **Speeding** data is 48.75. The standard deviation for this sample is 5.62.
   a. The value of 44 is below the mean and will have a negative z-score: $z = (44 - 48.75)/5.62 = \mathbf{-.85}$
   b. The value of 59 is above the mean and will have a positive z-score: $z = (59 - 48.75)/5.62 = \mathbf{1.82}$

Z-scores for data usually fall between -2 and 2 with most of the data concentrated close to $z = 0$.

1   A listing of the highest points of elevation for each of the 50 states is given below. Find the largest and smallest observations, the upper and lower quartiles, and the median. Then draw the box-and-whisker plot for the data.

Highest elevations in states

State	Max Ht.	State	Max Ht.	State	Max Ht.	State	Max Ht.
Fla	345	Minn	2301	W Va	4861	Idaho	12662
Del	442	Conn	2380	Okla	4973	Mont	12799
La	535	Ala	2405	Me	5267	Nev	13140
Miss	806	Ark	2753	N Y	5344	N M	13161
R I	812	Pa	3213	Neb	5426	Utah	13728
Ill	1235	Md	3360	Va	5729	Hawaii	13796
Ind	1257	Mass	3487	N H	6288	Wyo	13804
Ohio	1549	N D	3506	Tenn	6634	Wash	14410
Iowa	1670	S C	3560	N C	6684	Colo	14433
Mo	1772	Kan	4039	S D	7242	Calif	14494
N J	1803	Ky	4139	Tex	8749	Alaska	20320
Wis	1951	Vt	4393	Ore	11239		
Mich	1979	Ga	4685	Ariz	12633		

☺ 2   Box-and-whisker plots are often used to compare two sets of data. Draw the box-and-whisker plots for the following data and compare them.

Data set A: The beginning salary of two-year college graduates in civil engineering and in business administration one year was studied. The data appears below.

A. Beginning Salary Data

	Minimum	Q_L	Median	Q_U	Maximum
Civil Engineers	14 400	15 500	16 000	17 800	19 000
Business Administration	12 000	14 500	15 500	16 750	17 500

Data set B: The Academy Awards for Best Actress and Best Actor were first given in 1928. The table below shows the ages of the winners in each category between the years 1948 and 1990. Draw box-and-whisker plots of these observations. What do you conclude?

Age of Academy Award Winners for Best Actress and Best Actor

	34	41	27	24	48	27
	40	26	34	37	24	60
	61	28	30	26	42	35
Actress	31	28	26	33	38	21
	41	35	45	26	41	
	33	33	31	30	34	
	38	33	34	37	41	
	26	38	74	49	61	
	56	40	47	51	30	41
	43	51	43	46	56	44
	42	38	37	48	56	60
Actor	32	35	40	39	45	61
	33	62	52	35	38	
	41	42	31	39	41	
	49	56	40	48	38	
	43	38	76	55	35	

☺3.   Given the mean and standard deviation of a set of data, find the z-score for the data in question (x).

a.   Mean: 45   b.   Mean: 66
      Standard Deviation   5   Standard Deviation   7.5
      x   38   x   80

b.   Mean: 4.24   d.   Mean: 13130
      Standard Deviation   .015   Standard Deviation   1033
      x   4.255   x   12345

4.   Find the standard deviation of the data and then find the z-score for each value.

Z-score	X	$x - \bar{x}$	$(x - \bar{x})^2$
	5		
	6		
	8		
	10		
	11		
	12		
	13		
	15		
	$\bar{x} =$		$s =$

5.4   Linear Models of Best Fit

According to an article in the *News and Observer* December 11, 1995, "spending on Medicaid in North Carolina has risen dramatically over the past decade." The reporter attributed this increase largely to the "state and federal actions to expand the number of poor, elderly and disabled people eligible for benefits." He also attributed the increase in spending to "the state's growing elderly population and medical inflation."

Medicaid Spending in N C	
Year	Dollars *(in billions)*
1986	.78
1990	1.65
1992	2.08
1994	3.00

Medicaid Spending

Objectives of this section. The student will be able to
- *construct a scatter plot for data*
- *determine if a linear model might fit the data*
- *find and test a linear model on a small number of points*

Regression

The process of using one variable to predict another is called *regression*. For example, if you know a man's height you can make a good prediction about his weight. As another example, if you know the number of pages in a book, you could make a reasonable estimate about the weight of the book. Of course, when estimates like these are made, there will be errors. Errors may occur because other factors are not considered in the model.

The data that is to be examined is usually graphed as a *scatter plot--* the graph of all the ordered pairs. The *model* that will be used to *relate* the two variables is a prediction equation, sometimes known as a *regression equation*.

When two variables are related, there may be a *regression equation* that can be used to *model* the physical relationship with an algebraic relationship.

5.41

Models

There are many ways to find the model for a data set. In the next few exercises, you will develop an idea of what the <u>best</u> model is.

Linear Models

Consider the information at the beginning of this section. A scatter plot of the data appears to the right. Since these four points *seem* to follow a *linear pattern*, we should use a *linear function* to model the relationship. Obviously, with only four points, it is difficult to tell if the relationship is linear or not linear. It is best to consider more than four points when trying to model a relationship. But for this purpose, a four point data set will be sufficient.

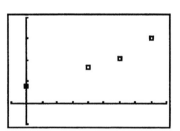

Follow the three steps outlined below to find the first *model* for these four ordered pairs.

Note

The equation of a line can be written in **point-slope form**: $y - y_1 = m (x - x_1)$ where m is the slope of the line, and (x_1, y_1) is a point on the line.

Example

The equation of the line with a slope of 3 passing through the point (1,2) is: $y - 2 = 3 (x - 1)$ which can be simplified: $y = 3x - 1$. This last equation is written in the **slope-intercept form**: $y = mx + b$.

STEP #1: SCATTER PLOT

Plot the data with *Years* on the x-axis. (To make the job easier, assign the year **1986** as $t = 0$). The y-axis will be in billions of *Dollars* and range from -1 to 4. (We start the y-axis at -1 to make the graph easier to read, even though it does not make sense to place -1 billions of dollars on the y-axis.)

STEP #2: FIND A MODEL FOR THE DATA

"Eyeball" these points to decide what kind of graph they are shaped like. Since they seem to be fairly linear, we need to find the equation of a line. We will choose two of the plotted points and write the equation of the line which passes through them. The equation of this line is a model that shows the relationship between time (in years) and dollars spent (in billions). The equation (model) should be written in functional notation like this:

Dollars Spent on Medicaid (time) $=$ *M (t)* $=$ *m t + b*

STEP #3: TEST THE MODEL

To analyze your model we will determine the <u>vertical</u> distance the model (line) is from each of the four points. The chart below will aid you in this determination.

Analyze the model's FIT Using Points (0, .78) and (4, 1.65)					
Actual Data		**Model Data** M(t) =		**Difference (D) and squared difference (D²) between the data and the model**	
Year	$ Spent	t	M(t)	D = $ Spent - M(t)	D²
0	.78	0			
4	1.65	4			
6	2.08	6			
8	3.00	8			
	SUM of the columns:				
	Largest Difference:				

In order to determine the line of "best fit," what would you consider as a measure of best fit? What are the different ways to measure "best fit?"

1.

2.

3.

STEP #4: FIND THE LINE OF BEST FIT

Because there are four points, there are SIX possible lines you could draw as a linear model. Repeat this process for the other five models that could be used with this data.

Analyze the model's FIT Using Points (0, .78) and (6, 2.08)					
Actual Data		**Model Data** M(t) =		**Difference (D) and squared difference (D²) between the data and the model**	
Year	$ Spent	t	M(t)	D = $ Spent - M(t)	D²
0	.78	0			
4	1.65	4			
6	2.08	6			
8	3.00	8			
	SUM of the columns:				
	Largest Difference:				

Analyze the model's FIT
Using Points (0, .78) and (8, 3.00)

Actual Data		Model Data M(t) =		Difference (D) and squared difference (D²) between the data and the model	
Year	$ Spent	t	M(t)	D = $ Spent - M(t)	D²
0	.78	0			
4	1.65	4			
6	2.08	6			
8	3.00	8			

SUM of the columns:

Largest Difference:

Analyze the model's FIT
Using Points (4, 1.65) and (6, 2.08)

Actual Data		Model Data M(t) =		Difference (D) and squared difference (D²) between the data and the model	
Year	$ Spent	t	M(t)	D = $ Spent - M(t)	D²
0	.78	0			
4	1.65	4			
6	2.08	6			
8	3.00	8			

SUM of the columns:

Largest Difference:

Analyze the model's FIT
Using Points (4, 1.65) and (8, 3.00)

Actual Data		Model Data M(t) =		Difference (D) and squared difference (D²) between the data and the model	
Year	$ Spent	t	M(t)	D = $ Spent - M(t)	D²
0	.78	0			
4	1.65	4			
6	2.08	6			
8	3.00	8			

SUM of the columns:

Largest Difference:

Analyze the model's FIT
Using Points (6, 2.08) and (8, 3.00)

Actual Data		Model Data $M(t) =$		Difference (D) and squared difference (D^2) between the data and the model	
Year	$ Spent	t	M(t)	D = $ Spent - M(t)	D^2
0	.78	0			
4	1.65	4			
6	2.08	6			
8	3.00	8			
	SUM of the columns:				
	Largest Difference:				

After the instructor completes the first table with the class, five groups will work on the other tables. The data from each group will be shared with other groups.

Which of the six lines would you choose as your model, the line of best fit? Why?

☺1. Use the steps listed above to find *"the best model"* for the six data points listed below. There are 15 possible models. __DO NOT__ attempt to try all 15 models. Examine the scatter plot and try only three, trying to find the best model from these three attempts.
Data Points:

Point A: (0, .78)   Point D: (7, 2.40)

Point B: (4, 1.65)   Point E: (8, 3.00)

Point C: (6, 2.08)   Point F: (9, 3.50)

☺2. Write a paragraph that defines the "line of best fit."

5.5   Help! Using Spreadsheets for Lines of Best Fit

Spreadsheets

Use a spreadsheet to find the least-squares line for the given data. Plot the data and the least-squares line on the same axis. How good is the fit?

Template

The following *template* may be provided by your instructor. A template is a spreadsheet in which the *formulas* have already been set up to provide the desired results. In this template, you only need to change the data in the first two columns <u>and</u> define the independent variable, dependent variable, and output location for the regression analysis.

Medicaid Spending

The formulas in columns C, D, and E all begin with a equal sign. Some spreadsheets require this or another sign to signify a formula instead of a literal expression. The *C-column* uses the regression output to calculate predictions for y based on the *coefficient of x* and the *constant* from the regression output. The *error* and *squared error* aren't really needed, but are nice to see. To graph this information, use *markers* only for the *B-column* and *line* for the *C-column*.

	A	B	C	D	E	F	G
1	Actual Data		Model (Prediction)	Error	Sq Error	Regression Output	
2	x	y	y'	$y - y'$	$(y - y')^2$	Constant	-22.1309
3	86	.78	=$G$7*A3+$G$2	=B3-C3	=D3^2	Std Err of Y Est	.216907
4	90	1.65	=$G$7*A4+$G$2	=B4-C4	=D4^2	R Squared	.963204
5	92	2.08	=$G$7*A5+$G$2	=B5-C5	=D5^2	No. of Observations	4
6	94	3.00	=$G$7*A6+$G$2	=B6-C6	=D6^2	Degrees of Freedom	2
7						x-coefficient	0.265286
8						Std Error of Coeff.	0.036664

Examples

Below, you will find two examples of regression analysis and the accompanying graphs.

D A T A		MODEL	ERROR	SQ ERROR		Regression Output:	
- - - -		-----	-----	--------	Constant		-3.0635
x	y	y'	y - y'	(y-y')^2	Std Err of Y Est		1.00166
4	0	0.84	-0.84	0.71	R Squared		0.98939
5	1	1.82	-0.82	0.67	No. of Observations		15
7	3	3.77	-0.77	0.60	Degrees of Freedom		13
9	6	5.73	0.27	0.07			
12	8	8.66	-0.66	0.43	X Coefficient(s)	0.97674	
13	10	9.63	0.37	0.13	Std Err of Coef.	0.02805	
15	12	11.59	0.41	0.17			
17	15	13.54	1.46	2.13			
18	16	14.52	1.48	2.20	MODEL:		
19	17	15.49	1.51	2.27	y =	0.98 x +	-3.0635
22	19	18.42	0.58	0.33			
23	18	19.40	-1.40	1.96	L I N E A R R E G R E S S I O N		
30	26	26.24	-0.24	0.06			
31	27	27.22	-0.22	0.05			
35	30	31.12	-1.12	1.26			
			-0.00	13.04			

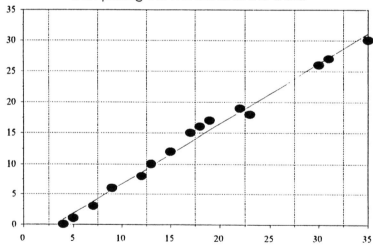

Linear Regression
Comparing the Model and the Data

DATA		MODEL	ERROR	SQ ERROR
- - - - - -		- - - - - - -	- - - - - - -	- - - - - - - - -
x	y	y'	y - y'	(y-y')^2
7	11	10.99	0.01	0.00
9	14	14.89	-0.89	0.80
12	20	20.75	-0.75	0.57
13	23	22.71	0.29	0.09
14.5	25	25.64	-0.64	0.40
15	30	26.61	3.39	11.48
16	30	28.57	1.43	2.06
18	33	32.47	0.53	0.28
19.2	34	34.81	-0.81	0.66
20	35	36.38	-1.38	1.90
21	37	38.33	-1.33	1.77
22	39	40.28	-1.28	1.65
23	41	42.24	-1.24	1.53
24	44	44.19	-0.19	0.04
25	49	46.14	2.86	8.17
S U M:			-0.00	31.38

Regression Output:

Constant	-2.68248
Std Err of Y Est	1.55367
R Squared	0.980795
No. of Observations	15
Degrees of Freedom	13
X Coefficient(s)	1.95298493
Std Err of Coef.	0.07579532

MODEL:

y = 1.95 x + -2.68

LINEAR REGRESSION

y' is the best guess of y for any x
using the model shown above.

Linear Regression
Comparing the Data and the Model

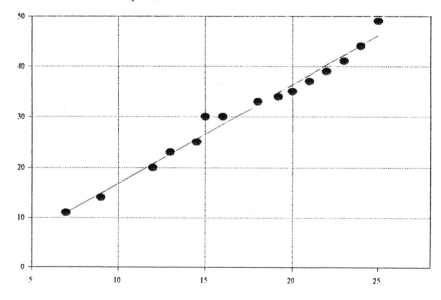

1. In a research project to determine the amount of drug which remains in the bloodstream after a given dosage, the amounts y (in mg of drug/dL of blood) were recorded after t hours.

Amount of drug remaining after Time						
t, time (hours)	1.0	2.0	4.0	8.0	10.0	12.0
a, amount (mg/dL)	7.6	7.2	6.1	3.8	2.9	2.0

Use your model to find the amount of drug that should be present at $t = 3.0$ hours, at $t = 13.0$ hours.

☺2. The record time for running the mile is shown in the table to the right. Find the least-squares regression line for the data.

What does your model predict that the record time will be in 2000?

Are there any flaws in the model?

Record times for running the mile				
Year	Time (sec)		Year	Time (sec)
1880	263.2		1945	241.4
1882	261.4		1954	239.4
1884	259.4		1957	238.0
1894	258.2		1958	237.2
1895	257.0		1962	234.5
1911	255.4		1964	234.1
1913	254.6		1965	233.6
1915	252.6		1966	231.3
1923	250.4		1967	231.1
1931	249.2		1975	229.4
1933	247.6		1979	229.0
1934	246.8		1980	228.8
1937	246.4		1981	228.4
1942	246.2		1985	226.3
1943	244.6			
1944	241.6			

3   The following table is a sample of some past annual mean salaries for teachers in elementary and secondary schools, along with the annual per capita beer consumption (in gallons) for Americans. Find the model that relates the two sets of data.

Teacher Salary and Beer Consumption

Year	1960	1965	1970	1972	1973	1983
Mean Teacher Salary	$5000	$6200	$8600	$9700	$10200	$16400
Per Capita Beer Consumption (gal)	24.02	25.46	28.55	29.43	29.68	35.2

What conclusions can you make with this data?

4   The following table shows the average SAT-math and SAT-verbal scores of all individuals that took the test in each of 12 states. Find the least squares regression line for the data.

SAT-Math and SAT-Verbal Scores in 12 States

	Twelve State - Average Score											
SAT-Math	410	485	440	420	430	400	390	400	450	430	410	420
SAT-Verbal	370	445	405	382	390	360	350	355	410	400	390	380

What conclusions can you make with this data?

☺   Place the data from your "odometer readings" project onto a spreadsheet. Then,
   1.   construct a scatter plot of the data.
   2.   find the linear regression model for the data.
   3.   predict the number of miles you will drive in a year based on your linear model.

You should place a printout of the spreadsheet, model and prediction in your portfolio. If it is possible, place a printout of your scatter plot there as well.

Chapter 6   Applications of Linear Equations

Although algebra appears in many situations, often people do not even realize it! In this chapter we will examine some fundamental concepts in algebra that are useful tools to solve a wide array of problems.

6.1   Systems of Linear Equations

Objectives of this section. The student will be able to

- *identify a solution to a system of linear equations*
- *find the approximate solution to a system of equations using a graph of the system*

The picture to the right displays two relationships between *bricks*, *prisms*, and a weight. Since both examples represent "balanced" beams, the combination of weights on the left side of the beam is equal to the weight on the right side of the beam. How much does each brick and each prism weigh?

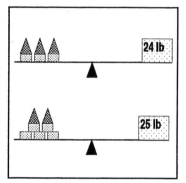

Concrete
to
Abstract

Systems of Equations

A *linear equation* is one that can be written in the form

$$ax + by = c.$$

An *ordered pair* that is a *solution* to a linear equation is a value of x and a value of y that, when substituted into the equation, makes the equation true. A pair of linear equations is called a *system of two linear equations*. A *solution* to a system of linear equations is any ordered pair that makes <u>both</u> equations in the system true.

A pair of linear equations is called a system of linear equations.

$$\begin{cases} 2x - 3y = 1 \\ x + 2y = 11 \end{cases}$$

A solution to a system of linear equations is an ordered pair that makes both equations in the system true: (5,3)

6.11

A solution of a system of Linear Equations

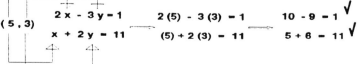

Check!

The Number of Solutions to a System of Two Linear Equations

When two linear equations are graphed, one of three situations may occur.

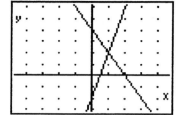

1. One Solution

If the lines are different and intersect, the equations are *independent* and the system is *consistent*. **One solution exists.** The slope of the lines must be different.

2. Infinitely Many Solutions

If the lines *coincide,* the equations are *dependent* and the system is *consistent*. **Infinitely many solutions exist.** The slope of the two lines must be the same.

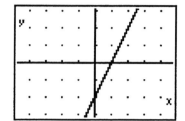

3. No Solution

If the lines are different and *parallel*, the equations are *independent* and the system is *inconsistent*. **No solution exists.** The slope of the two lines must be the same.

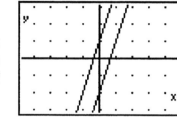

When a system is consistent and has one solution, that solution can be found graphically. Consider the two equations shown in the figure boxes below as a system of two linear equations in x and y.

By plotting points to graph each equation, the ordered pair that is common to both lines can be found. A table of Y_1 and Y_2 is shown.

In this case, the solution is found as the tables are built. The ordered pair (1,2) makes both equations true. This is obvious in the table since when x = 1, both Y_1 and Y_2 equal 2.

Graphs and Tables

With the aid of a graphing calculator, the solution can be found by *tracing* and *zooming* toward the point of intersection.

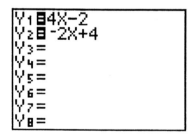
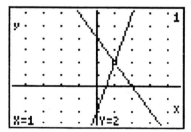

X	Y1	Y2
-1.000	-6.000	6.0000
0.0000	-2.000	4.0000
1.0000	2.0000	2.0000
2.0000	6.0000	0.0000
3.0000	10.000	-2.000
4.0000	14.000	-4.000
5.0000	18.000	-6.000

X=1

If the solution is not composed of two integers, the likelihood of finding the exact solution in a table is remote. Consider the next example.

The system made up of the two equations shown in the figure box to the right has one solution. The three figures shown below indicate three steps in the process of finding an *approximation* to that solution. The first graph represents the graph of each equation shown in a *standard window*. The next two graphs indicate a "closer look" at the point of intersection. This is done by *zooming*.

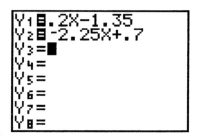

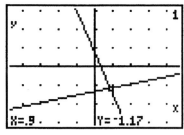

Standard Window

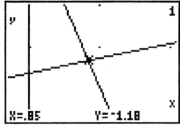

Zoom In

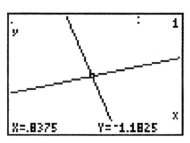

Zoom In Again

Exact vs. Approximate Solutions

You can see from the figure to the right that this solution, (**.8375, -1.1825**), is an approximate solution. It is not the exact solution since it does not make <u>both</u> equations in the system true. To find the <u>exact</u> solution, different procedures must be employed.

```
.2(.8375)-1.35
                -1.1825
-2.25(.8375)+.7
                -1.1844
```

Group Work

Determine if the given ordered pair **is** or **is not** a solution to the equation.

1. $(3, 5)$ $y = 3x - 4$ | 4. $(.5, .25)$ $y = 2x - 1$

2. $(1.25, 4)$ $y = 4x - 1$ | 5. $(-3, 5)$ $y = .45x + 3.5$

3. $(1.3, -2)$ $2x - .5y = 3.75$ | 6. $(-3, -2)$ $2x - 3y = 12$

Determine if the given ordered pair **is** or **is not** a solution of the given system of two equations.

7. $(.25, 3)$ $2x + .5y = 2$
$4x - 2y = -5$

8. $(.35, -.5)$ $y = 2x - 1.2$
$y = x - .925$

9. $(0, -2)$ $y = 4x - 2$
$y = 3x + 2$

10. $(4.625, -1.375)$ $3x + 5y = 7$
$2x + 6y = 1$

The next two sets of pictures provide you with (1) a system of two linear equations, (2) the graph of the two equations, and (3) the window for the graph. The solution of each system is made up of integers. Estimate the solution from the graph and <u>verify</u> the solution by substituting the ordered pair into <u>both</u> equations in the system.

1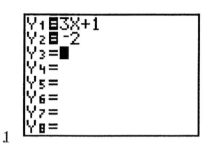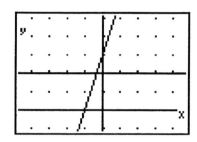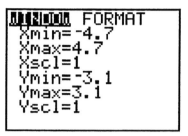

2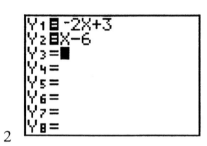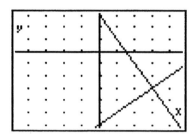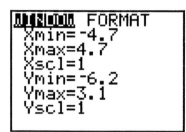

☺ 1. In these exercises, tell whether the ordered pair is a solution of the system of equations.

a. (3, 6) $4x - 2y = 0$
              $x + 2y = 15$

b. (-2, 4) $\frac{1}{2}x - 2y = -9$
               $x + \frac{3}{4}y = 1$

c. (0, $\frac{1}{2}$) $7.2x + 2y = 1$
                 $5x + 4y = 2$

d. (1.3, 2.7) $2x - 5y = -10.9$
                 $x + y = 5$

e. (5.3, -3.5)
        $2.7x - 7.2y = 35.91$
        $5x + 2y = 19.5$

f. (-2, -3) $-4x - 2y = 14$
               $-5x + 2y = 4$

☺ 2. Estimate the solution of the system of linear equations from the graph provided. After arriving at an estimate, check the ordered pair in both equations to see if the result is an approximate solution or an exact solution.

a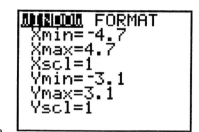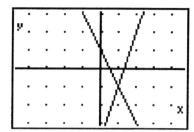

b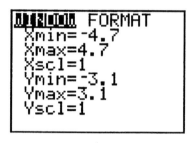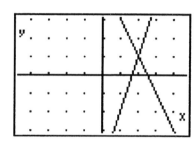

c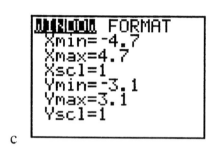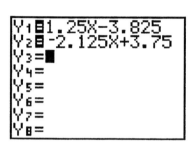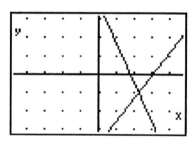

d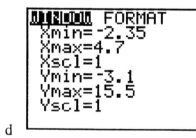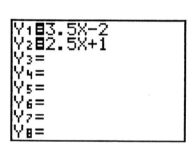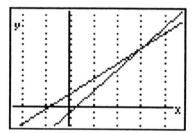

3. All types of systems may be solved using this graphing approach. Find the solutions of the following system using the graphs provided. Show that the ordered pairs <u>are</u> solutions.

a.

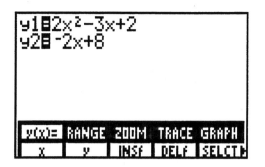

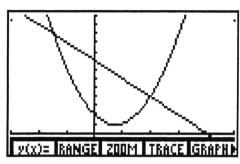

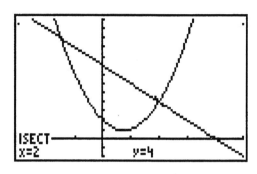

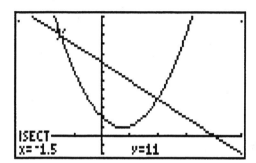

b.

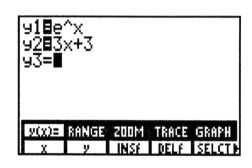

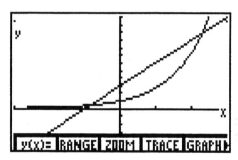

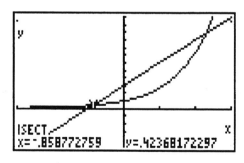

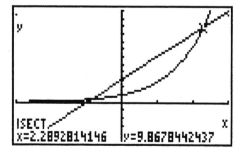

6.2   Solving Systems of Linear Equations - Algebraic Procedures

Objectives of this section.   The student will be able to

- *find the solution to a system of two linear equations using algebraic techniques*

Square D has two machines that are used to make plastic molds for electric starters. The set up for a certain mold is $400 on the small machine and $950 on the large machine. Once set up, the unit cost for the smaller machine is $3 and $2 for the larger. The *break-even* point is the number of units, x, that need to be manufactured to make the cost the same using either machine.

Solving Systems of Two Linear Equations

Algebraic procedures may be employed to solve systems of linear equations when technology is not available or if technology cannot be used. The methods fall into two categories: the *substitution* method or the *elimination* method.

Consider the system of equations shown to the right.

$$\begin{cases} x = 2 \\ 2x - 3y = 8 \end{cases}$$

Substitution

The first equation identifies that 2 is the value of x in the· solution of the system. If that is so, then because of the definition of a solution of a system, the value of x in the second equation is also 2. Therefore, the procedure is to *substitute* 2 for x in the second equation and solve for y. The result is (2,-4/3) as the solution to the system.

$$2(2) - 3y = 8$$
$$4 - 3y = 8$$
$$-4 = 3y$$
$$\frac{-4}{3} = y$$
$$\left(2, -\frac{4}{3} \right)$$

The procedure is the same for the second system shown here. The first equation identifies the "value" of x (five less than twice y). This equation is said to be *solved for x in terms of y*. When the information in the first equation is added to the information in the second equation, the solution can be found.

$$\begin{cases} x = 2y - 5 \\ 2x - 3y = -7 \end{cases}$$

Since x is 2y - 5 (from the first equation), then the x in the second equation is also 2y - 5. The substitution is made and the solution found as shown.

$$2(2y - 5) - 3y = -7$$
$$4y - 10 - 3y = -7$$
$$y - 10 = -7$$
$$y = 3$$
$$x = 2y - 5$$
$$x = 2(3) - 5$$
$$x = 6 - 5$$
$$x = 1$$
$$(1, 3)$$

Recall that a solution to a system of two linear equations makes each equation true. A solution can and *should* be checked.

Check: $1 = 2(3) - 5 = 6 - 5 = 1$
$2(1) - 3(3) = -7 \quad 2 - 9 = -7 \quad -7 = -7$

Solve this system using the substitution method

$$y = 3x + 8$$
$$6x - y = -7$$

The substitution method can be employed even when neither of the equations is solved for a variable. The first step is to solve one of the equations for one of the variables.

$$2x + 3y = 8$$
$$2x = 8 - 3y$$
$$\frac{1}{2}(2x) = \frac{1}{2}(8 - 3y)$$
$$x = 4 - \frac{3}{2}y$$

Solving for x

The two examples to the right provide a review of the process of solving for a variable.

$$4x - 2y = 9$$
$$4x = 2y + 9$$
$$4x - 9 = 2y$$
$$\frac{1}{2}(4x - 9) = \frac{1}{2}(2y)$$
$$2x - 4.5 = y$$

Solving for y

Here are two more examples of the substitution method.

$2x + 3y = 8$	$x = 4 - 1.5y$	$7(4 - 1.5y) - 3y = 1$	$2x + 3(2) = 8$	
$7x - 3y = 1$	$7x - 3y = 1$	$28 - 10.5y - 3y = 1$	$2x + 6 = 8$	$(1, 2)$
		$-13.5y = -27$	$2x = 2$	
		$y = 2$	$x = 1$	Solution
System	Solve for x	Substitution	Completing the Ordered Pair	

$3x - 4y = 8$	$3x - 4y = 8$	$3x - 4(2x - 4.5) = 8$	$3(2) - 4y = 8$	
$4x - 2y = 9$	$y = 2x - 4.5$	$3x - 8x + 18 = 8$	$6 - 4y = 8$	$(2, -\frac{1}{2})$
		$-5x = -10$	$-2 = 4y$	
		$x = 2$	$-\frac{1}{2} = y$	Solution
System	Solve for y	Substitution	Completing the Ordered Pair	

Applying the Concept

A company decides to market its "Flyer" shoes for $4 more than the "Glider" shoes. Let F be the price of the Flyers and G be the price of the Gliders. It also decides, based on production costs, that 5 Flyers and 8 Gliders must sell for $800. Write a system of equations and solve using the substitution method.

Elimination

In the other algebraic method used to solve systems of equations, the equations are combined in such a way as to eliminate all the terms involving one of the variables. This process makes use of a property of equality that allows equals to be added to equals and get equals. This property is illustrated to the right.

$$A = B$$
$$C = D$$
$$\text{-----------}$$
$$A + C = B + D$$

Employing this property on a system of equations can eliminate all the terms involving one variable if the coefficients are opposite in value.

$$-3x + 7y = 9$$
$$3x + 3y = 1$$
$$\text{-----------}$$
$$10y = 10$$
$$y = 1$$

As an example, consider the x terms in the system shown. When the two equations are added together, the result is an equation in one variable. That equation can be solved and the result used to find the solution to the system.

$$3x + 3(1) = 1$$
$$3x + 3 = 1$$
$$3x = -2$$
$$x = -\frac{2}{3}$$

$$(-\frac{2}{3}, 1)$$

If the coefficients are not opposites in value, another property of equality can be used to make them opposite. In the example below, both sides of the second equation must be multiplied by three in order to make the coefficient of y in that equation opposite in value to the coefficient of y in the first equation.

$$IF \quad A = B$$
$$THEN \quad AC = BC$$

$2x - 9y = 1$	$2x - 9y = 1$	$2x - 9y = 1$	$11x = 55$
			$x = 5$
$3x + 3y = 18$	$3(3x + 3y) = 3(18)$	$9x + 9y = 54$	$2(5) - 9y = 1$
			$-9y = -9$
			$y = 1$
System	Multiply Both Sides of One Equation	Simplify	$(5, 1)$
			Solved

In this case, both equations are changed to use the elimination method.

$$3x - 7y = 1 \qquad -4(3x - 7y) = -4(1) \qquad -12x + 28y = -4 \qquad \begin{array}{l} 55y = -55 \\ y = -1 \\ 3x - 7(-1) = 1 \end{array}$$

$$4x + 9y = -17 \qquad 3(4x + 9y) = 3(-17) \qquad 12x + 27y = -51 \qquad \begin{array}{l} 3x = -6 \\ x = -2 \\ (-2, -1) \end{array}$$

System   Multiply Both Sides   Simplify   Solved
of Each Equation

Group Work

Solve: $3x + 5y = 1$
$6x + 9y = 3$

Two other important points:

1.   Recall that not all systems of two linear equations have one solution. If the algebraic methods produce a statement that is <u>false</u>, the system is *inconsistent*. If the algebraic methods produce a statement that is always true, the system is *consistent*, but *dependent*.

$$\begin{cases} 2x + 4y = 12 \\ x = 6 - 2y \end{cases} \qquad\qquad \begin{cases} 5x - 7y = 11 \\ -5x + 7y = 3 \end{cases}$$
$$2(6 - 2y) + 4y = 12$$
$$12 = 12 \qquad\qquad\qquad\qquad 0 = 14$$

Consistent and Dependent   Inconsistent

2.   Either of these procedures can be used on any system of two linear equations, but usually one method works better than the other. It all depends on how the equations are arranged.

Group Work

Choose either algebraic method to solve the system.

1. $2x - 54 = 1$
$3x + 24 = 11$

2. $x = 24 + 7$
$2x - 3y = 13$

Applications

Applications involving systems of two linear equations provide enough information for two equations.

Example

One taxi company charges **$1.50 + $.25** per *half-mile*. Another charges **$2.00 + $.20** per *half-mile*. For what length trip are the two charges the same?

The variables in this problem are:

h = number of half-miles driven

C = charge

The two equations are shown and the substitution method employed. The two charges would be the same on a trip of 5 miles.

$$\begin{cases} C = 1.50 + .25h \\ C = 2.00 + .20h \end{cases}$$

$$2.00 + .20h = 1.50 + .25h$$
$$.50 = .05h$$
$$10 = h$$

$$(10, 4.00)$$

When solving problems in which the equations are written from a verbal description of the problem, it is most important to **define** the variables <u>before</u> writing any equations.

Example

A radio station approached a company with two packages for placing adds on its station. The first package contained five thirty-second adds and ten sixty-second adds. It cost $4500. The second package cost $4500, but it contained twelve thirty-second adds and six sixty-second adds. How much did the different length adds cost individually?

Let

t = price of a thirty-second ad

m = price of a sixty-second ad

The two equations are shown and the elimination method employed. The 30-second ad cost $200 and the minute ad cost $350.

$$\begin{cases} 5t + 10m = 4500 \\ 12t + 6m = 4500 \end{cases}$$

$$3(5t + 10m) = 3(4500)$$
$$-5(12t + 6m) = -5(4500)$$

$$15t + 30m = 13500$$
$$-60t - 30m = -22500$$
$$-45t = -9000$$

$$t = \$200$$
$$m = \$350$$

The answer to a problem stated verbally should be a verbal statement. Notice that once the numbers have been found that satisfy the mathematics, a sentence is written to answer the verbal question.

---

1. Solve the following systems of equations.

$$\begin{cases} y = 3x + 5 \\ 3x + 2y = 9 \end{cases} \qquad \begin{cases} y = 3 - 5x \\ 3x - 5y = 9 \end{cases}$$

$$\begin{cases} 3x + 5y = 3 \\ 3x + 2y = 9 \end{cases} \qquad \begin{cases} 2x - y = 3 \\ 7x + 2y = 5 \end{cases}$$

☺2. Linda is on a business trip and needs to rent a car for a day. The prices at two agencies are different, and she wants to rent the less expensive car.

Agency A: $10 per day plus $.50 per mile
Agency B: $40 per day plus $.25 per mile

The decision is based on the number of miles Linda intends to drive. Let c be the cost (in dollars) of the rental and let m be the number of miles driven. Write the two equations that correspond to each Agency's cost. Use algebra to solve for the number of miles at which point the costs are the same for each agency.

3. A homeowner needs to dig a well for his new home. He calls two well drilling companies and gets two quotes.

Digger 1: $400 plus $5 per foot
Digger 2: $250 plus $6 per foot

Write equations for each company [$C(f)$]. Find the number of feet at which the two costs will be the same.

4. Two businesses compared their payoff on a new computer.

Business 1:   Paid $4800 for the computer and is paying $400 per month.
Business 2:   Paid $6000 for the computer and is paying $600 per month.

After how many months will the two businesses owe the same amount on their computer? (Note: there are two answers.)

5. A flatbed truck made two hauls. It hauled contaminated barrels of kerosene and barrels of cleaning fluid. The first load tipped the scales at 8700 pounds and contained 6 barrels of kerosene and 10 barrels of cleaning fluid. The second load contained 9 barrels of each liquid and weighed 8550 pounds. How much did a barrel of each liquid weigh?

6. Running a machine usually involves set up costs and unit costs. Set up costs include the cost of preparing a machine to do a certain job. Unit costs depend on the number units being manufactured. Use the information in the introductory vignette to find the break-even point for the two machines described.

☺7. A company builds two kinds of wooden children's chairs: standard and deluxe. The per unit manufacturing costs are shown in the table. The company has budgeted $960 for cost of materials and $1450 for labor. How many of each type does it plan to build?

Model	Cost of Materials	Cost of Labor
Standard	24	30
Deluxe	30	50

6.3 Multiple Approaches to Systems of Linear Equations

Objectives of this section. The student will be able to
- *produce a numerical table of values in order to compare the two y-values in the system*
- *determine between what two x-values a solution to the system occurs*
- *numerically zoom-in on a solution*
- *construct a spreadsheet that will allow numerical investigations of systems of equations*

A landscape architect was experimenting with a design and changed the dimensions of a patio. Originally, the patio was square, but the experimental design was six feet less on one side and nine feet more on the other. The area of the experimental patio was the same as the original one.

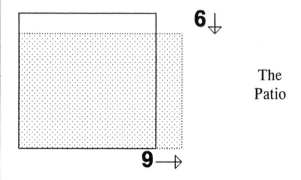

The Patio

Numerical Investigations

Algebraic and graphical methods for solving systems of equations are necessary. It is also important to recognize that a numerical investigation into a system of equations is a valid method of finding the solution. The spreadsheet has made this method even more inviting.

Begin with two tables which contain ordered pairs of two different, unknown functions. Is (2.5,3.49) a solution to the first equation?

Is (2.5,3.49) a solution to the second equation?

FUNCTION 1		FUNCTION 2	
Input	Output	Input	Output
1	1.65	1	1
1.5	2.12	1.5	1.84
2	2.72	2	2.82
2.5	3.49	2.5	3.95
3	4.48	3	5.20
3.5	5.75	3.5	6.55
4	7.39	4	8
4.5	9.49	4.5	9.55
5	12.1	5	11.1

The approximate location of any solutions can be found by closely examining the table.

Consider x = 1.5
    the output for function 1 is *more* than the output for function 2.
Consider x = 2
    the output for function 1 is *less* than the output for function 2.

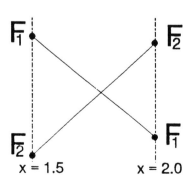

This change in the *relative position* of the two y-values indicates a point of intersection, as the graph clearly demonstrates.

x	Y_1	Y_2	
0	1	0	Y_1 is larger than Y_2
.5	1.28	.35	Y_1 is larger than Y_2
1	1.65	1	Y_1 is larger than Y_2
1.5	2.12	1.84	Y_1 is larger than Y_2
2	2.72	2.83	Y_1 is smaller than Y_2

A table can provide a *closer* look by reducing the increment in x.

Again, there is a change in the relative position of the y-values between x = 1.85 and x = 1.90.

X	Y1	Y2
1.6	2.2255	2.0239
1.65	2.2819	2.1195
1.7	2.3396	2.2165
1.75	2.3989	2.315
1.8	2.4596	2.415
1.85	2.5219	2.5163
1.9	2.5857	2.619

X=1.85

x	Y_1	Y_2	
1.85	2.5219	2.5163	Y_1 is larger than Y_2
1.90	2.5857	2.619	Y_1 is smaller than Y_2

Another table provides even a closer look. One more numeric-zoom provides a close approximation to the solution.

(1.8571, 2.5308)

X	Y1	Y2
1.853	2.5257	2.5224
1.854	2.5269	2.5244
1.855	2.5282	2.5265
1.856	2.5294	2.5285
1.857	2.5307	2.5306
1.858	2.532	2.5326
1.859	2.5332	2.5347

X=1.857

X	Y1	Y2
1.857	2.5307	2.5306
1.8571	2.5308	2.5308
1.8572	2.531	2.531
1.8573	2.5311	2.5312
1.8574	2.5312	2.5314
1.8575	2.5313	2.5316
1.8576	2.5315	2.5318

X=1.8571

Spreadsheets

A spreadsheet template that can easily be used to find solutions of any two equations using this numeric-zoom process is shown below.

	A	B	C	D	E	F
1	0	Initial x		x	Function 1	Function 2
2	1	Increment		+A1	(D2)^(1.5)	(2.71828)^(.5*D2)
3				+D2+$A$2	(D3)^(1.5)	(2.71828)^(.5*D3)
4				+D3+$A$2	(D4)^(1.5)	(2.71828)^(.5*D4)
5				+D4+$A$2	(D5)^(1.5)	(2.71828)^(.5*D5)

The initial x and the increment can be changed by entering new values in A1 or A2 respectively. The two functions are written in E2 and F2 and copied vertically.

Group Work

Use the two "calculator displays" below to find out between what two x-values a solution occurs?

X	Y1	Y2
1	0	.30103
1.1	.11	.34242
1.2	.24	.38021
1.3	.39	.41497
1.4	.56	.44716
1.5	.75	.47712
1.6	.96	.50515

X=1

X	Y1	Y2
.4	1.6	1.4832
.5	1.625	1.5811
.6	1.65	1.6733
.7	1.675	1.7607
.8	1.7	1.8439
.9	1.725	1.9235
1	1.75	2

X=1

Equilibrium

Demand and *supply* are two important concepts in economics. In simple terms, demand measures the quantity of items consumers are willing to purchase and supply measures the quantity of items manufacturers are willing to produce. Many variables might affect supply and demand, but selling price is a major factor.

The *equilibrium point* is the intersection of the demand and supply curves. Remember the chair company from a previous homework? Suppose the supply function for the deluxe model is

S(p) = 2p - 160

This function indicates that the company is willing to produce *S(p)* chairs when the chairs sell for a price, *p*. Since S(80) = 2(80) - 160 = **0**, the company is not willing to produce any chairs for a selling price of $80, or less. How many chairs are they willing to produce for a selling price of

$100? _____

$120 _____

The demand for the chairs is also a function of the price:

D(p) = 120 - p

This function indicates that the demand is zero if the price reaches $120. How many chairs would be bought at a selling price of $100? _____
$80 _____

X	Y1	Y2
80	0	40
85	10	35
90	20	30
95	30	25
100	40	20
105	50	15
110	60	10

X=80

These two functions may be analyzed numerically. Let $Y_1 = S(p)$ and $Y_2 = D(p)$ and $x = p$. Use the table at the right to isolate the equilibrium point.

X	Y1	Y2
90	20	30
91	22	29
92	24	28
93	26	27
94	28	26
95	30	25
96	32	24

X=90

This numeric process provides a very accurate picture of the solution of two equations. A graph can then be used with the process to find the first estimate of the solution. The graph below shows that the equilibrium point is close to 93.

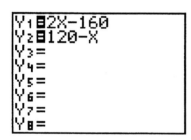 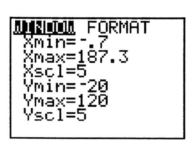 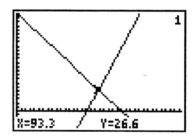

Practice

☺ 1. Construct a spreadsheet which shows the solution to the following problems. If you cannot print your spreadsheet out for your portfolio, then just write the two lines that isolate the solution onto your page.

A)   A homeowner needs to dig a well for his new home. He calls two well-drilling companies and gets two quotes.
    Digger 1: $400 plus $5 per foot
    Digger 2: $250 plus $6 per foot

Write equations for each company [C(f)]. Find the number of feet at which the two costs will be the same.

B) Two businesses compared their payoff on a new computer.

Business 1: Paid $4800 for the computer and is paying $400 per month.

Business 2: Paid $6000 for the computer and is paying $600 per month.

After how many months will the two business owe the same amount on their computer? (Caution, there are two answers).

2. A company producing bicycle reflectors has fixed costs of $2100 and variable costs of $.80 per reflector. The selling price of each reflector is $1.50. Use a spreadsheet to numerically find the break-even point?

3. The supply curve for a new software product is x = 2.5p - 500, and the demand curve for the same product is x = 200 - .5p.
   a. At a $250 selling price, how many items would be supplied? How many would be demanded?
   b. At what price would no items be supplied? demanded?
   c. What is the equilibrium price for this product?
   d. How many units would be produced and demanded at the equilibrium price?

4. After considerable study a business that sells Chicago Cub baseball hats finds that its supply curve and demand curve can be modeled for a range in the selling price between $4.00 and $9.00.

   Supply as a function of the price: $S(p) = 1.2p^3 - 4.9p^2$

   Demand as a function of the price: $D(p) = 177p - 18.8p^2$

   Find the Equilibrium point using a spreadsheet.

5. Use the graph to help isolate the solution. Then, use a table to further isolate the more exact solution.

a.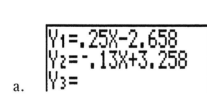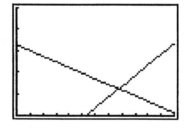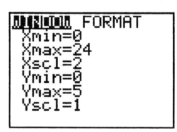

b.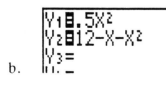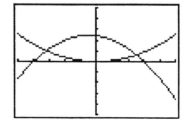

6.4 Matrices - An Introduction

Objectives of this section. The student will be able to

- identify the elements of a matrix using row × column format
- perform addition and multiplication of matrices by hand
- perform matrix operations using a spreadsheet

	Product		
	X	Y	Z
Labor	20	25	28
Material	10	18	14

Rows & Columns of Data

Matrices

In economics it is often convenient to examine data. Many situations in economics as well as in other disciplines often lead to the study of a rectangular array of numbers. The data is divided into classes which are designated by rows and columns. The figure above is such an example. You have already seen that systems of linear equations arise in many applications. This chapter will eventually use matrices to solve systems of equations with several variables. This type of system appears in applications in many areas.

The mathematical term for a rectangular array of numbers enclosed with brackets is a *matrix*. The numbers in each matrix are called its *elements*. The rows and columns are numbered successively beginning at the top with the first row and on the left with the first column. Each element is described by its location; by what row and columns it is in. This is done by using a *subscript*. The element $m_{2 \times 3}$ would be in the 2nd row and the third column of the matrix *M*.

A *matrix* is a rectangular array of numbers, called *elements*. The *dimension* of the matrix is the number of rows and the number of columns. The matrix shown below has 3 rows and 2 columns and has dimension *3 × 2*.

$$M_{3 \times 2} = \begin{bmatrix} m_{1 \times 1} & m_{1 \times 2} \\ m_{2 \times 1} & m_{2 \times 2} \\ m_{3 \times 1} & m_{3 \times 2} \end{bmatrix}$$

6.41

Group Work

Consider the matrix shown to the right. What are the dimensions of the matrix? _____

Using subscripts, identify the two elements that have a value of zero. _____ _____

$$R = \begin{bmatrix} 1 & 3 & 5 & 7 \\ 6 & 2 & 9 & 0 \\ 3 & 0 & 4 & 5 \end{bmatrix}$$

Equality of Matrices

Two matrices are *equal* if the elements in the same location are equal. That is, matrix **A** is equal to matrix **B** if and only if every $a_{i \times j} = b_{i \times j}$.

The two matrices shown here are equal. That means that x = 3, y = 4 and z = 5.

$$\begin{bmatrix} x \\ y \\ z \end{bmatrix} = \begin{bmatrix} 3 \\ 4 \\ 5 \end{bmatrix}$$

Matrix Operations

This section has provided the basic facts about matrices. Addition and multiplication of matrices will now be introduced by example.

A company keeps track of the number of contacts each sales representative makes each month. The company also keeps track of the number of contracts each sales representative produces. This information can be written in matrix form as shown.

January's Totals

	Number of Contacts	Number of Contracts
Adam	23	11
Bill	34	13
Carolynn	27	9
Denise	17	12

February's Totals

	Number of Contacts	Number of Contracts
Adam	16	6
Bill	24	5
Carolynn	21	6
Denise	13	7

The two matrices have the same dimension (*4 × 2*), but are not equal. It would make sense that the company would want to add the information from January and February together for a subtotal. It would also make sense to add Adam's January contacts to Adam's February contacts; and that is exactly how matrix addition takes place. The elements in the *same location* are added together: $j_{1 \times 1} + f_{1 \times 1} = s_{1 \times 1}$.

$$J = \begin{bmatrix} 23 & 11 \\ 34 & 13 \\ 27 & 9 \\ 17 & 12 \end{bmatrix}$$

$$F = \begin{bmatrix} 16 & 6 \\ 24 & 5 \\ 21 & 6 \\ 13 & 7 \end{bmatrix}$$

Matrix addition <u>requires</u> that the two matrices have the **same dimension**. An example of matrix addition is shown below.

$$\begin{bmatrix} 2 & 2 & 3 \\ 5 & 9 & 1 \end{bmatrix} + \begin{bmatrix} 5 & 3 & 0 \\ 3 & 0 & 1 \end{bmatrix} = \begin{bmatrix} 7 & 5 & 3 \\ 8 & 9 & 2 \end{bmatrix}$$

Add the two matrices **J** and **F** together.
Explain the meaning of this new matrix.

To subtract two matrices, simply add the opposite of each element. This process is just like regular addition: **3 - 5 = 3 + (-5)**.

Matrix Multiplication

Next, consider matrix multiplication. You may not feel that the definition of matrix multiplication is realistic, or natural, but as the concept becomes clearer you should be convinced that the definition is appropriate and extremely practical for applications.

> ### Matrix Multiplication
> Let **A** be an $m \times n$ matrix and **B** be an $n \times p$ matrix. The product of the two matrices **A * B,** is the matrix $\mathbf{C}_{m \times p}$ whose entry $c_{i \times j}$ is obtained as follows: take the sum of the products formed by multiplying, in order, each entry in row i of **A** by the corresponding entry in column j of **B**.

6.42

There are three important points we must remember about matrix multiplication.
1.   The two matrices must be *conformable*. That is, the number of columns of **A** must be equal to the number of rows of **B**.
2.   The product of two matrices is a third matrix with the same number of rows as **A** and the same number of columns as **B**.
3.   Matrix multiplication is done in a specific order. **A * B** may not necessarily be equal to **B * A**.

The following graphics may help explain the concept.

$$\begin{bmatrix} 6 & 0 & 2 \\ 1 & 4 & 5 \end{bmatrix} * \begin{bmatrix} 3 & 7 & -1 \\ 3 & 1 & 6 \\ 0 & -1 & 0 \end{bmatrix} = \begin{bmatrix} 18 & 40 & -6 \\ 15 & 6 & 23 \end{bmatrix}$$

$(6)(-1) + (0)(6) + (2)(0) = -6 + 0 + 0 = -6$

$$\begin{bmatrix} 6 & 0 & 2 \\ 1 & 4 & 5 \end{bmatrix} * \begin{bmatrix} 3 & 7 & -1 \\ 3 & 1 & 6 \\ 0 & -1 & 0 \end{bmatrix} = \begin{bmatrix} 18 & 40 & -6 \\ 15 & 6 & 23 \end{bmatrix}$$

$(1)(3) + (4)(3) + (5)(0) = 3 + 12 + 0 = 15$

Use the matrices shown to perform the indicated operations. If the operation can not be performed, indicate that the two matrices are not conformable to multiplication.

1. **A × B**   2. **C × B**   3. **B × C**   4. **C × A**

$$A = \begin{bmatrix} 1 & 0 & 0 \\ 0 & 1 & 0 \\ 0 & 0 & 1 \end{bmatrix} \quad B = \begin{bmatrix} 1 & 2 & 0 \\ 7 & 3 & 1 \\ 1 & 0 & 1 \end{bmatrix} \quad C = \begin{bmatrix} 1 & 2 & 4 \\ 4 & 2 & 0 \end{bmatrix}$$

The Identity Matrix

After doing the above multiplication, it is evident that the matrix **A** is special. It is called the *identity matrix*. It acts like the "one" in regular multiplication. The identity matrix must be square. It is made up of "1" on the diagonal and "0" for all other elements.

Using Spreadsheets

Matrix multiplication may involve many calculations. To perform multiplication on large matrices or ones with non-integral elements, you may wish to employ technology. The spreadsheet does matrix multiplication easily. Simply identify the two matrices, identify the output and "go." Calculators may also do matrix multiplication. After you have done a few with pencil and paper, please feel free to use the technology that is available to do the rest.

Practice

☺1.   Construct a matrix **A** so that:
the matrix is 3 × 3, and
$a_{1 \times 1} = 4, a_{2 \times 1} = 2, a_{2 \times 2} = 3 , a_{3 \times 3} = 4,$
and the sum of each row is 9, **and**
the sum of each column is 9.

2.   Use the three matrices shown below to perform the indicated operation. If the operation cannot be done, state why.
a. **A + C**   b. **B + C**   c. **A × B**
d. **B × A**   e. **A × C**

$$A = \begin{bmatrix} 3 & 4 & 5 \\ 2 & 1 & 0 \end{bmatrix} \quad B = \begin{bmatrix} 1 & 2 \\ 0 & 1 \\ 0 & 1 \end{bmatrix} \quad C = \begin{bmatrix} 1 & 2 & 4 \\ 4 & 2 & 0 \end{bmatrix}$$

3. Use a matrix to summarize the number of copier units sold by Adam, Bill and Carolynn during the last six months of 1994 **(S94)** and 1993 **(S93)** . Use a matrix to show the monthly increase or decrease for each person each month **(S94 - S93)**.

Number of copier units sold per month (1994)

	July	August	Sept	Oct	Nov	Dec
Adam	4	5	8	4	3	1
Bill	3	3	6	3	2	2
Carolynn	2	8	8	2	0	1

Number of copier units sold per month (1993)

	July	August	Sept	Oct	Nov	Dec
Adam	2	2	3	2	3	1
Bill	3	3	4	1	3	1
Carolynn	0	3	4	5	2	3

The cost of a copier changed each month. The table below shows the selling price of a copier for each month in 1994 and 1993.

Selling price for copiers

	July	Aug	Sept	Oct	Nov	Dec
1993	$550	575	600	600	625	650
1994	$725	725	750	775	775	800

Multiply the 1993 matrix **(S93$_{3\times6}$)** by the price matrix **(P$_{6x1}$)** matrix:

$$\begin{bmatrix} 550 \\ 577 \\ 600 \\ 600 \\ 625 \\ 650 \end{bmatrix}$$

The result of this product is a 3 x 1 matrix that represents the total production (the total amount of sales) for each person during the last six months of 1993.

Do the same for the 1994 sales.

☺4. The statistics for several players during a baseball game appear below. Hint: multiply each player's statistics (a 1 × 7 matrix) by the 7 × 1 matrix made up of the "pay."

Game Summary (fictional)

Player	At Bats	Runs	Hits	Home runs	RBIs	Errors	Ks
Mantle	5	2	2	1	4	1	1
Mays	4	1	3	1	1	0	0
Marris	5	1	1	0	0	1	2
Kimball	5	3	3	2	6	1	2

Some people have argued that baseball players should be paid based on their performance. Use matrix multiplication to find the "pay" for each player if an at bat is worth $100, a run is $150, a hit is $200, a home run $500, an RBI is $200, an error is -$150, and a strikeout (K) is -$200.

5. A developer builds a housing complex featuring two-, three-, and four-bedroom units. Each unit comes in two different floor plans. The matrix **P** (for production) tells the number of each type of unit for this development.

Building Plans
Plan I   Plan II

$$P = \begin{bmatrix} 10 & 5 \\ 25 & 10 \\ 15 & 10 \end{bmatrix} \begin{matrix} \text{2 bedrooms} \\ \text{3 bedrooms} \\ \text{4 bedrooms} \end{matrix}$$

Many materials are used in building these homes, but this model will be simplified to include only lumber, concrete, fixtures, and labor. The matrix **M** (for materials) gives the amounts of these materials used (in appropriate units of each).

Building Plans

	Lumber	Concrete	Fixtures	Labor	
M =	7	8	9	20	Plan I
	8	9	9	22	Plan II

Find the matrix product **P × M** which gives the amount of material needed for the development.

Let **A = P × M.**

If the cost of each unit of material is given by the matrix **C** (for cost), then the total cost of each model for this development is **A × C**.

Cost per unit

$$C = \begin{bmatrix} 1800 \\ 190 \\ 2000 \\ 2000 \end{bmatrix} \begin{matrix} \text{Lumber} \\ \text{Concrete} \\ \text{Fixtures} \\ \text{Labor} \end{matrix}$$

6. Use the information in the box at the beginning of this section to construct a **3 × 2 matrix** with *Labor* and *Material* on **top**. Multiply that matrix by the cost matrix shown here to find the total cost for each item. How much will each item cost in labor and materials? Write a paragraph on matrix multiplication that explains why a row should be multiplied by a column. Use this example in your explanation.

$$C = \begin{bmatrix} 14 \\ 8 \end{bmatrix}$$

6.5 Matrices and Their Inverse

Objectives of this section. The student will be able to
- *identify and construct an identity matrix of any dimension*
- *perform matrix operations to test to see if two matrices are inverses*
- *construct a matrix equation from a system of two linear equations*
- *use matrix methods to solve a system of linear equations*

$$\begin{bmatrix} -2 & 1 & 0 \\ 1.5 & -.5 & -.5 \\ 2 & -1 & 1 \end{bmatrix} * \begin{bmatrix} 2 & 2 & 1 \\ 5 & 4 & 2 \\ 1 & 0 & 1 \end{bmatrix} = \begin{bmatrix} 1 & 0 & 0 \\ 0 & 1 & 0 \\ 0 & 0 & 1 \end{bmatrix}$$

The Identity Matrix

Matrix Equations

This course has included several types of equations and the process used to solve them. The equation to the right is not made up of regular variables; each letter represents a matrix. Thus it is a *matrix equation* and the solution is a matrix. This section will consider a matrix such as **C** in the second equation. Notice that the product of **C** and **A** is **I**, the identity matrix. Because of this, the matrix **C** is said to be the *inverse of A: A^{-1}*. Inverses are used to solve matrix equations much like the reciprocal is used in regular algebra. The method for finding them, though, is not as easy. Technology will be used to find the inverse of a square matrix.

$$A \cdot X = B$$
$$C \cdot A \cdot X = C \cdot B$$
$$I \cdot X = C \cdot B$$
$$X = C \cdot B$$

or

$$A \cdot X = B$$
$$A^{-1} \cdot A \cdot X = A^{-1} \cdot B$$
$$I \cdot X = A^{-1} \cdot B$$
$$X = A^{-1} \cdot B$$

The Inverse of a Matrix

The spreadsheet and programmable, graphing calculators are capable of finding the inverse of a matrix. In a spreadsheet, simply identify the matrix and ask for its inverse. Some matrices don't have inverses; those that aren't square and those that are *singular* are two such examples. If a square matrix does not have an inverse, it is called a singular matrix.

The matrix A^{-1} is the inverse of the square matrix A if and only if $A * A^{-1} = A^{-1} * A = I$

6.51

Group Work

Find the product of the two matrices and indicate whether they are or are not inverses of one another.

Is **A** the inverse of **B**? $A = \begin{bmatrix} 6 & 5 \\ 7 & 6 \end{bmatrix}$ $B = \begin{bmatrix} 6 & -5 \\ -7 & 6 \end{bmatrix}$

Is **C** the inverse of **D**? $C = \begin{bmatrix} 2 & 5 & 2 \\ 1 & 3 & 1 \\ 1 & 5 & 11 \end{bmatrix}$ $D = \begin{bmatrix} 2.8 & -4.5 & -.1 \\ -1 & 2 & 0 \\ .2 & -.5 & .1 \end{bmatrix}$

As the name implies, the *coefficient matrix* is made of elements which are coefficients of variables in an equation. When each equation in a system of equations is arranged in order such as

$$ax + by + \ldots = k$$

then the coefficient matrix can be taken directly from the coefficients as shown below. The matrix associated with the matrix **B** in the matrix equation is also shown. It is sometimes called the *constant matrix*.

System: $\qquad$ *Coefficient Matrix:* $\qquad$ *B:*

$$\begin{cases} 2x + 3y - 4z = 7 \\ 5x - y + z = 3 \\ 7x + 2z = 9 \end{cases} \qquad \begin{bmatrix} 2 & 3 & -4 \\ 5 & -1 & 1 \\ 7 & 0 & 2 \end{bmatrix} \qquad \begin{bmatrix} 7 \\ 3 \\ 9 \end{bmatrix}$$

We now have all the ingredients to use matrices to solve a system of equations.

Solving a System of Equations Using Matrices

In order to solve the system shown, first write a matrix equation. The matrix equation is made up of the coefficient matrix, the variable matrix and the constant matrix. Notice that if the two matrices on the left-hand side of the equation are multiplied together, the result would be three expressions exactly like the three left-hand sides of the three equations in the system.

$$\begin{cases} 2x + 3y - 4z = 7 \\ 5x - y + z = 3 \\ 7x \qquad + 2z = 9 \end{cases}$$

$$\begin{bmatrix} 2 & 3 & -4 \\ 5 & -1 & 1 \\ 7 & 0 & 2 \end{bmatrix} * \begin{bmatrix} x \\ y \\ z \end{bmatrix} = \begin{bmatrix} 7 \\ 3 \\ 9 \end{bmatrix}$$

$$\begin{bmatrix} .04878 & .14634 & .02439 \\ .07317 & -.78049 & .53659 \\ -.17073 & -.51220 & .41463 \end{bmatrix} * \begin{bmatrix} 2 & 3 & -4 \\ 5 & -1 & 1 \\ 7 & 0 & 2 \end{bmatrix} * \begin{bmatrix} x \\ y \\ z \end{bmatrix} = \begin{bmatrix} .04878 & .14634 & .02439 \\ .07317 & -.78049 & .53659 \\ -.17073 & -.51220 & .41463 \end{bmatrix} * \begin{bmatrix} 7 \\ 3 \\ 9 \end{bmatrix}$$

Next, find the inverse of the coefficient matrix. This is done using technology. The inverse matrix is placed to the **left** of A and to the **left** of C. Since matrix multiplication is <u>not</u> commutative, this is very important. The multiplication is carried out and rounded appropriately. The result on the left side is, as expected, the identity matrix. The right side is a 3 × 1 matrix.

$$\begin{bmatrix} 1 & 0 & 0 \\ 0 & 1 & 0 \\ 0 & 0 & 1 \end{bmatrix} * \begin{bmatrix} x \\ y \\ z \end{bmatrix} = \begin{bmatrix} 1 \\ 3 \\ 1 \end{bmatrix}$$

$$\begin{bmatrix} x \\ y \\ z \end{bmatrix} = \begin{bmatrix} 1 \\ 3 \\ 1 \end{bmatrix}$$

After multiplying by the identity matrix, (we really do not since we know it is the *identity matrix)* the result is a very simple equality. The only way these two matrices can be equal is if each corresponding element is equal. Therefore, the solution is obvious.

$$x = 1$$

$$y = 3$$

$$z = 1$$

Checking the result in the original system proves that the solution is indeed correct.

Important: The inverse matrix is placed to the <u>left</u> of A and to the <u>left</u> of C since matrix multiplication is not commutative.

$$\begin{cases} 2(1) + 3(3) - 4(1) = 7 \\ 5(1) - (3) + (1) = 3 \\ 7(1) \qquad + 2(1) = 9 \end{cases}$$

1. Solve the following system of equations. Use the information provided in the box at the beginning of the section.

$$\begin{cases} 2x + 2y + z = 5 \\ 5x + 4y + 2z = 9 \\ x + z = 4 \end{cases}$$

2. a. Find A^{-1}, and multiply $A^{-1} \times A$

$$A = \begin{bmatrix} 3 & 4 & -2 \\ 6 & 9 & 0 \\ 7 & -6 & 1 \end{bmatrix}$$

b. Find A^{-1}, and multiply $A \times A^{-1}$

$$A = \begin{bmatrix} 3 & 4 & -2 \\ 6 & 8 & -6 \\ 7 & -8 & 6 \end{bmatrix}$$

☺ 3. Solve the following systems using matrix methods. Your portfolio should show not only the solution but the inverse of the coefficient matrix.

$$\begin{cases} x + 2y + z = 1 \\ x + 3y + 4z = 6 \\ 2x + 4y + 3z = 7 \end{cases} \qquad \begin{cases} 2x + 3y - z = 7 \\ 2x + y - z = 5 \\ 2x + 4y + 3z = 12 \end{cases}$$

6.6   Linear Programming-Inequalities

Objectives of this section. The student will be able to
- *graph the solution of a linear inequality*
- *graph the solution of a system of linear inequalities*
- *find the corner points of the solution of a system of linear inequalities*

Linear Programming

One of the most difficult tasks in a real-world problem is the construction of a mathematical model that simulates the situation. In the 1940s a new mathematical model called *linear programming* was found to be applicable to a wide range of situations in which the maximum or minimum value in some application was needed. This section will introduce you to linear programming. Before proceeding, a review of some basic concepts might prove useful.

The graph of the line $y = 3x + 1$ is shown to the right. Recall that the points on the line are the ordered pairs (x, y) that make the equation true. An *inequality* is an equation in which the equal sign, $=$, has been replaced by either of these symbols: less than, $<$; greater than, $>$; less than or equal to, $\leq$; or greater than or equal to, $\geq$. Solutions of inequalities are ordered pairs that make the inequality true. The solution set of an inequality usually consists of an area bounded by the solution of the equation. The solution of the inequality $y \geq 3x + 1$ is the region shaded in figure to the right. It consists of all the ordered pairs for which the y value is greater than or equal to three times x plus one. It would be impossible to list all these ordered pairs--just like it is impossible to list the ordered pairs that make an equation true. But you don't have to. It makes sense that all the ordered pairs that make y *greater than or equal to* 3x + 1 are those ordered pairs on one <u>side</u> of the line that makes y *equal* to 3x + 1. So the steps involved in graphing an inequality are:

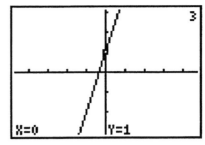

Equation: y = 3x + 1

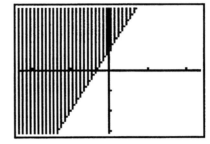

Inequality: y ≥ 3x + 1

1. Graph the equation.
2. Test a point on either side of the line.
3. If the point makes the inequality true, then shade that area.
4. If the point does not make the inequality true, then shade the other side.

Here is an example: Graph $2x + 3y \geq 5$

Make a
Table of Values

x	y
0	5/3
2.5	0
4	-1

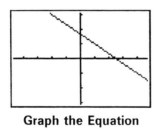

Graph the Equation

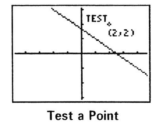

Test a Point

(2,2) Works! Shade it.
$2(2) + 3(2) \geq 5$

The solution of an inequality containing ≥ or ≤ is a half-plane <u>and</u> the line. This is noted by using a **solid** line. The solution of an inequality containing < or > is <u>only</u> the half-plane. This is noted by using a <u>dashed</u> line. Here are some other examples.

Graph: *y < 3*

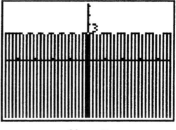

Y < 3

The graph of y = 3 is a horizontal line through y = 3. If you test the point (0, 0) on one side of the line, it makes the inequality true, *0 < 3*. So that side of the line is shaded. The line should be dashed.

Graph: *x ≥ 1*

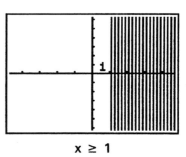

x ≥ 1

The graph of x = 1 is a vertical line through x = 1. If you test a point (0, 0) on one side of the line, it does not make the inequality true *0 ≯ 1*. So the other side of the line is shaded. The line should be solid.

Graph: *y < 2x + 5*

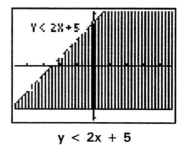

y < 2x + 5

The graph of y = 2x + 5 is a line through (0,5) with a slope of 2. If you test a point (0, 0) on one side of the line, it makes the equation true *0 < 2(0) + 5*. So that side of the line is shaded, and the line should be dashed.

Group Work

Graph the solution to the following inequalities.

1.   y ≥ 0
2.   x ≥ 0
3.   2x + 3y ≤ 18

The Solution Set of a System of Linear Inequalities

The solution set of a *system of linear inequalities* consists of all the points whose coordinates <u>simultaneously</u> satisfy all of the inequalities. Geometrically, it is the region which is <u>common</u> to all the regions determined by the inequalities in the system. It is the **intersection** of all the regions which are solutions of each inequality in the system. Here is an example.

To find the solution of this system, simply graph the solution of each inequality and then put them together, **shading <u>only</u>** the region that is common to all the solutions. If you choose to graph all the equations on one axis, you may want to use dashed lines

$$\begin{cases} 3x + 5y \le 150 \\ x > 10 \\ y \ge x - 5 \end{cases}$$

to begin with until you are certain what is to be solid. Instead of shading as you go, you can initially draw little arrows originating on the line designating which side of the line is to be shaded and shade as you "see" the intersection.

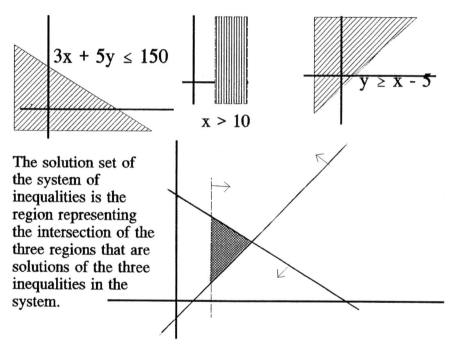

The solution set of the system of inequalities is the region representing the intersection of the three regions that are solutions of the three inequalities in the system.

The shaded region in the figure above contains many points. But if you select any point in that region and substitute appropriately in any inequality in the system, that point makes the inequality true. For example, the point *(12, 10)* is located within the shaded region. As you can see, it does make each inequality true.

$$\begin{cases} 3(12) + 5(10) \le 150 \\ (12) > 10 \\ (10) \ge (12) - 5 \end{cases}$$

> A *solution of a system of inequalities* contains all the points whose coordinates simultaneously satisfy all of the given inequalities.

6.61

Divide into groups of three. Each person will take one of the inequalities in the system and find the solution. Once everyone in the group has found the solution to their inequality, put the solutions together to find the solution to the system of linear inequalities.

$$\begin{cases} 3x + 5y \le 30 \\ x - 2y > -1 \\ x \ge 0 \end{cases} \qquad \begin{cases} x + 5y \le 6 \\ y > -4x + 5 \\ y \ge 0 \end{cases}$$

Finding the Coordinates of the Corners of the Shaded Region

It should be fairly obvious that the *corners* of the shaded region that denotes the solution of a system of linear inequalities are the solutions of various *pairs* of linear **equations**. That is, the corners are made from two intersecting lines. The point where two lines intersect is the solution of a system of two linear equations.

Consider the example previously used. The first two equations make up a system of equations that can be solved by substitution: *(10, 24)*.

$$\begin{cases} 3x + 5y \le 150 \\ x > 10 \\ y \ge x - 5 \end{cases}$$

It is important to organize your work so that you know which two equations to place in a system and which corner you are finding the coordinates of.

The system made up of the second and third equations can be solved using substitution: *(10, 5)*.

The system composed of the first and third equations was solved using technology: *(21.875, 16.875)*

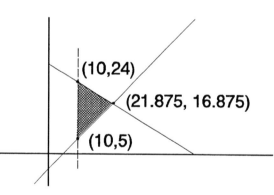
(10,24)
(21.875, 16.875)
(10,5)

This enclosed region has corners on all sides. The solution to a system of inequalities does not necessarily have corners on all edges. The solution may be unbounded on one edge and have no corners.

Group Work

Find the corners of each region that makes up the solution for each system in the group work above.

Find the solution for each inequality. Also, find the coordinates of the corners to the region. Each corner is the solution to a system of <u>two</u> linear equations.

1. $\begin{cases} 5x + 2y \geq 10 \\ \quad x - 2y \leq 2 \end{cases}$

2. $\begin{cases} x + 2y \geq 10 \\ \quad x > 0 \end{cases}$

☺3. $\begin{cases} 3x + 2y \geq 12 \\ 2x - 2y \leq 3 \\ \quad\quad y > 0 \end{cases}$

☺4. $\begin{cases} 5x + 4y \geq 30 \\ \quad\quad y \leq 2x + 1 \\ \quad\quad x > 0 \end{cases}$

6.7   Linear Programming - Introduction

Objectives of this section. The student will be able to

- *define the two main components of a linear programming problem: constraints and the objective function*
- *determine if a "solution" to a linear programming problem is feasible*
- *evaluate feasible solution based on the objective function*

Linear Programming Problems

Systems of linear inequalities provide the foundation for the study of an area of applied mathematics known as *linear programming*. Whenever questions like "How can I maximize the profit?" or "How can I minimize costs?" appear in problems, linear programming is usually a tool to help find an answer. Problems of this nature usually are quite complicated, containing many variables. These are beyond the scope of this course. But for problems containing only two variables, the process of isolating the feasible solutions and determining the best answer can be solved with the mathematical tools developed in this text.

Constraints

The solutions to such problems usually depend on *constraints*. These are usually limited resources that must be considered. In business, those resources may be labor or materials. In other applications, it could be vitamins or minerals. When the constraints are considered, only some answers are feasible. Those solutions that satisfy all the constraints make up the *feasible region*. The objective of a linear programming problem is usually to maximize or minimize something. In business, the cost of doing business may be dependent on several variables and within some constraints. In medicine, a diet must maximize the effect of several types of vitamins within given constraints.

The Objective Function

The solution to a linear programming problem is that ordered pair in the feasible region that either maximizes or minimizes the *objective function*. This process is sometimes referred to as *optimization*.

To get a better idea of such a problem, consider the following scenario.

A farmer has 100 acres on which to plant two crops: corn and wheat. To produce these crops, the farmer has certain expenses which are shown in the table.

After the harvest, the farmer must usually store the crops while awaiting favorable market conditions. Each acre yields 110 bushels of corn or 30 bushels of wheat. The limitations of resources are

Available capital: $15 000
Available storage facilities: 4 000 bushels

Allocation of Resources in Production	
Item	**Cost Per Acre**
Corn	
Seed	$12
Fertilizer	$58
Planting/care/harvesting	$50
Total	$120
Wheat	
Seed	$40
Fertilizer	$80
Planting/care/harvesting	$90
Total	$210

If the net profit per bushel of corn is $1.30 and for wheat is $2.00, how should the farmer plant the 100 acres to maximize the profits?

Working Through Some "Solutions"

In order to understand the problem more fully, possible solutions to the problem will be tested for feasibility.

Try planting all 100 acres in wheat. (100,0)
- The production is $30 \times 100 = 3000$ bushels, for a net profit of $3000 \times \$2 = \$6\,000$. But, to plant 100 acres with wheat would cost $\$210 \times 100 = \$21\,000$, and only $15\,000$ is available. This "solution" is **not feasible.**

Try planting 100 acres of corn. (0,100)
- The total cost is $\$120 \times 100 = \$12\,000$ and the net profit is $110 \times 100 \times \$1.30 = \$14\,300$. However, the yield of 11,000 bushels (110×100) cannot be stored since there are facilities to store only 4,000 bushels. This "solution" is **not feasible.**

Obviously, neither of the above scenarios are feasible. Try some other values and see if a maximum profit that is within the constraints of the problem can be achieved.

Try planting 50 acres of wheat and 50 acres of corn. (Just a wild guess!) (50,50)
- How much capital is needed?
  $50 * 120 + 50 * 210 = \$16,500$
  Since this "solution" does not satisfy one of the constraints, it is **not feasible.**

Thinking About the Results

Since wheat costs more to plant and the costs need to be reduced, try
planting 25 acres of wheat and 75 acres of corn. (We are working
under the assumption that we will maximize one of the constraints--
acreage. That may not be possible.)

(25,75)

○ How much capital is needed?

25 * $210 + 75 * $120 = $14,250

Since this "solution" satisfies the "capital constraint," we can
continue with the analysis.

○　　　 How much storage is needed?

25 * 30 + 75 * 110 = 9000 bushels

This is far beyond the available storage. Therefore, this "solution"
is **not feasible.**

This solution produces values within the amount of capital available,
but beyond the available storage facilities. We might consider
planting less than the available acreage.

Try planting 45 acres of wheat and 20 acres of corn.

(45,20)

○　　　 This scenario would cost: 45 * $210 + 20 * $120 =
$11,850.

○　　　 The storage needed is: 45 * 30 + 20 * 110 = 3550 bushels.

Since this satisfies <u>both</u> constraints, this **"solution" is feasible.** The profit can now be
calculated:

45 * 30 * $2.00 + 20 * 110 * $1.30 = $5560

Let us construct a table to summarize what we have done.

Group Work

Continue to analyze the problem by completing at least two other rows to the table with scenarios that <u>are</u> feasible. Discuss how one would know when a solution was optimal.

Available: 100 acres		Available: $15,000	Available: 4000	Maximize
Wheat	Corn	Cost	Bushels	Profit
100	0	21000	3000	n/a
0	100	14300	11000	n/a
25	75	14250	9000	n/a
45	20	11850	3550	5560

The Wadsworth Widget Company manufactures two types of widgets: regular and deluxe. Each widget is produced at a station consisting of a machine and a person who finishes the widgets by hand. The regular widget requires 3 hours of machine time and 2 hours of finishing time. The deluxe widget requires 2 hours of machine time and 4 hours of finishing time. The profit on the regular widget is $25 each, and on the deluxe widget, the profit is $30 each. If the workday allows for 100 hours of machine time and 120 hours of finishing time, how many of each type of widget should be produced at each station per day in order to maximize profit?

___

Group Work

Work on the Widget problem together. Be careful to fully understand the problem by reading it several times carefully. Discuss the problem within your group. The instructor should not have to explain the problem to the class. Often a table, like the one in the Farmer Brown example, is helpful in the analysis.

___

Practice

☺ Continue to work on the Widget problem for homework. Find at least two other feasible solutions in addition to the one(s) found in class.

Review the last two sections in order to be ready for the next section.

___

6.8   Linear Programming

Objectives of this section. The student will be able to

- *find the solution of a linear programming problem given the feasible region and the corner points*
- *write the objective function and the constraints for a linear programming problem*
- *solve a problem using linear programming*

Consider the region shown to the right. It is the solution to a system of linear inequalities:

$$y \leq -2x + 10$$
$$y \leq -3x + 14$$
$$y \geq 0$$
$$x \geq 0$$

Which point in that region makes the following expression the **greatest**?

$$30x + 40y$$

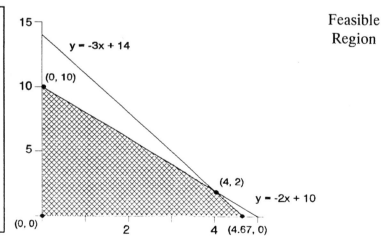

Feasible Region

___

The Optimal Solution

We will employ graphs to solve a linear programming problem. The constraints will be written as a system of linear inequalities. They will be graphed and the solution to the system will be the *feasible region*. Every point in the feasible region can not be tested in the *objective function*, but we can test each *corner point* of the region. With further analysis, it is apparent that the *optimal solution* will be a corner point.

> The *optimal solution* to a linear programming problem will be a corner point of the feasible region.

6.8.1

Summary

In summary, here are the steps to solve a linear programming problem.

> **Steps to solve a Linear Programming Problem (with two unknowns):**
> (1) Identify the unknowns (write a sentence to describe both unknowns).
> (2) Identify whether it is a maximization or minimization problem, and what is to be optimized.
> (3) Write the objective function.
> (4) Describe the constraints algebraically.
> (5) Find the feasible region graphically.
> (6) Test the corner points.
> (7) Find the solution.

6.8.2

Before tackling a problem from start to finish, examine the figures at the beginning of this section.

Group Work

Consider the figures at the beginning of this section. Make a table and test all the corner points of the feasible region in the objective function. Find at least two points on an edge (not a corner) and test those. Discuss why you think that only the corners can produce an optimal solution.

Farmer Brown

Go back to the Farmer Brown example and solve it using algebraic and graphical methods.

To formulate a **mathematical model**, begin by letting
  x = **number of acres** to be planted in corn
  y = **number of acres** to be planted in wheat

There are certain constraints. The number of acres planted cannot be negative, and the amount of available land is 100 acres or less. Because of storage and other limitations, all 100 acres may not be used). Therefore,
  $x \geq 0$
  $y \geq 0$
  $x + y \leq 100$

We also know that
$120 * x = $ expenses for planting corn
$210 * y = $ expenses for planting wheat

The total expenses cannot exceed $15,000. This is the available capital; therefore,
$$120x + 210y \leq 15,000$$

Also, the yields are
110 bushels * x = yield of acreage planted in corn
30 bushels * y = yield of acreage planted in wheat

The total yield cannot exceed the storage capacity of 4,000 bushels; therefore,
$$110x + 30y \leq 4,000$$

These constraints are summarized here
$x \geq 0$
$y \geq 0$
$x + y \leq 100$
$120x + 210y \leq 15,000$
$110x + 30y \leq 4,000$

Now, let P represent the total profit. The farmer wants to maximize this profit. A function that is to be maximized or minimized is called the **objective function.**

Profit from corn   = value * amount
= $1.30 per bushel * 110 bushels per acre * x acres
= $143 * x

Profit from wheat   = value * amount
= $2.00 per bushel * 30bushels per acres * y acres
= $60 * y

P = Profit from corn + profit from wheat
P = $143x + $60y

The Linear Programming Problem

The linear programming model is stated as follows:

Maximize: $P = \$143x + \$60y$
Subject to: $x \geq 0$
                  $y \geq 0$
                  $x + y \leq 100$
                  $120x + 210y \leq 15,000$
                  $110x + 30y \leq 4,000$

The feasible region and corner points are shown below.

Matrices were employed to find the solution of the last two inequalities. That process is shown in the figures to the right.

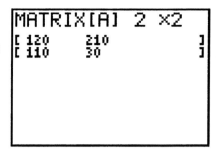

Construct Matrix A

Corner Points

(0, 71.43)
(20, 60)
(36.36, 0)
(0, 0)

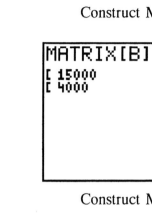

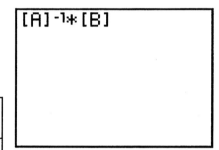

Construct Matrix B

A table is used to test the corner points.

Available: 100 acres		Available: $15,000	Available: 4000	Maximize
Corn	Wheat	Cost	Bushels	Profit
0	0	0	0	$0
0	71.43	$15000	2143	$4286
20	60	$15000	4000	$6460
36.36	0	$4363	4000	$5199

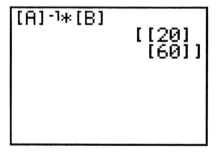

Multiply

Find the solution

The Solution

As you can see, the point (20, 60) is the *optimal solution* since it produces the maximum profit. It should not be concluded after this example that the optimal solution will be a corner point that is not on the axis. After all, if the price of wheat skyrocketed because of demand in Russia, it would not be inconceivable to plant all wheat.

Farmer Brown wanted to maximize his profit on 100 acres of land. He planned to plant wheat and corn. He was faced with two other constraints besides the land: capital and storage. The profit was maximized by planting 20 acres of corn and 60 acres of wheat for a profit of $6460.

Construct a linear programming model for the Wadsworth Widget Co. problem and solve it.

6.9   Linear Programming - More

Objectives of this section.   The student will be able to
■   *solve a linear programming problem*

Set up a linear programming model for each problem and find the optimal solution. Write the solution as part of a complete sentence within a paragraph that explains the constraints and identifies the limits of the solution.   Begin these in class and finish them for homework.   Your portfolio should clearly show the variables, a table that organizes the information, the constraints, a graph that shows the corner points, and an analysis of each corner point.

☺1.   Example: Marx Brothers Investment Co.

Marx Bros. is an investment company analyzing a pension fund for a certain company. A maximum of $10 million is available to invest in two places. Not more than $8 million can be invested in stocks yielding 12%, and at least $2 million must be invested in long-term bonds yielding 8%. The stock-to-bond investment ratio cannot be more than 1 to 3. How should Marx Bros. advise its client so that the pension fund will receive the maximum yearly return on investment?

[Note: to build this model you need to use the *simple interest formula:*
    $I = prt$]

2.   The Diet Problem
A convalescent hospital wishes to provide, at a minimum cost, a diet that has a minimum of 200 grams of carbohydrates, 100 grams of protein, and 120 grams of fats per day. These requirements can be met with two foods.

If food A cost $.29 per ounce and food B costs $.15 per ounce, how many ounces of each food should be purchased for each patient per day in order to meet the minimum requirements at the lowest cost?

Diet Problem			
	Contents per oz		
Food	Carbohydrates	Protein	Fats
A	10 g	2 g	3 g
B	5 g	5 g	4 g

☺ 3.   The Nutty Problem

A nut company sells two different mixtures of nuts. The cheaper mixture contains 80% peanuts and 20% walnuts, while the more expensive mixture contains 50% of each type of nut. Each week the company can obtain up to 1800 pounds of peanuts and up to 1200 pounds of walnuts from its sources of supply. How many pounds of each mixture should be produced in order to maximize profits if the profit is 10 cents from each pound of the cheaper mixture and 15 cents from each pound of the more expensive mix?

4.   The Two Machines Problem

A manufacturer makes two products, A and B, each of which requires time on three machines. Each unit of A requires 2 hours on the first machine, 4 hours on the second machine, and 3 hours on the third machine. The corresponding numbers for each unit of B are 5, 1, and 2, respectively. The company makes profits of $250 and $300 on each unit of A and B, respectively. If the numbers of machine hours available per month are 200, 240, and 190 for the first, second, and third machines, respectively, determine how many units of each product must be produced to maximize the total profit.

Chapter 7   Applications of Probability

This chapter will attempt to apply many of the concepts covered in the course. It will answer some questions that may have developed during this course about variation, sampling and error. When a model is supposed to do fairly well at predicting and does not, why doesn't it? Why is there more faith in a model made from many points than one that is made from just a couple? Before these questions can be fully answered, you must understand basic probability. This chapter will also consider the exponential function.

7.1 Introduction to Probability

Objectives of this section. The student will be able to
- *define probability*
- *determine the probability of an event in a sample space consisting on mutually exclusive and equally likely events*
- *determine the number of outcomes of an event*
- *determine the probability of an event from a histogram*

In the first part of this course, you made histograms showing the results of rolling one die several times and then rolling two dice several times. When one die is rolled, the results are either 1, 2, 3, 4, 5 or 6. The frequency for each result was always pretty much the same. When two dice are used the results are either 2, 3, 4, ..., 11, or 12. The frequency for a 2 or 12 was always less than 5, 6 or 7. Why were those two histograms so different?

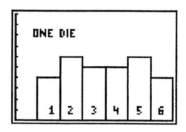

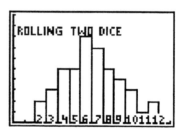

Experiments

A die is a good way to begin the study of probability. Each time a die is rolled, you are performing an *experiment*. An **experiment** is an observation of any physical occurrence. When you roll a die you expect to get one of six different results. These possible results make up a *sample space*. A **sample space** is the set of options or possibilities from which the outcomes of an experiment can be taken. The sample space for rolling one die is **S = {1, 2, 3, 4, 5, 6}**. The different possibilities in the sample space are *events*. An **event** is a subset of a sample space. For example, **1** is an event in the sample space for the experiment of rolling one die. The events in this sample space are *mutually exclusive*. That is, they can't happen at the same time. Two events are **mutually exclusive** if they may not occur simultaneously.

1. Let the experiment be the birth of one child. The sample space for this experiment is { **M, F** }. These two events are mutually exclusive.

2. Let the experiment be who comes to a party. The sample space for this event (those people invited to the party) might be: **{Ric, Lisa, Cyrus, Sue, Jay, Ted, Lynn, Beth, Ann}** These events are not mutually exclusive because even if Cyrus comes, the others may also.

Probability

The probability of an event is a number between zero and one that identifies the likelihood of an event happening. The closer the probability is to zero, the less likely the event is to happen. The closer the event is to one the more likely the event is to happen. Events with a probability of zero can't happen and events with a probability of one are certain to happen.

3-Day Forecast

30% Monday 70% Tuesday 0% Wednesday

Probabilities are usually expressed as decimals or percents. When the weather channel predicts is a 30% chance of rain, they are assigning a probability of .30 that at least .01 inch or rain will fall in a particular area. You can also find applications of probability in baseball or basketball. A batter's average predicts the likelihood of a hit. An average of .303 says that at the present time the batter has about a 30% chance of getting a hit. A basketball player's shooting percentage identifies the likelihood that the shooter will make a basket on the net shot. A shooting percentage of .445 means that the shooter has about a 45% chance of making the next shot.

We now turn to probabilities that can be computed.

> If an experiment has a sample space consisting of n mutually exclusive and equally likely events, then a *uniform probability model* assigns the probability of 1/n to each simple event.

7.11

Mutually Exclusive and Equally Likely

Since the experiment, rolling one die, has a sample space consisting of **six** mutually exclusive and equally likely events, then a uniform probability model assigns the probability of 1/6 to each simple event. The probability of rolling a <u>one</u> is 1/6 and the probability of rolling a <u>six</u> is 1/6.

If an experiment can occur in any of n $(n \geq 1)$ mutually exclusive and equally likely ways and if s of those ways are considered favorable to E, then the probability of the event E, denoted by $P(E)$, is

$$P(E) = \frac{s}{n} = \frac{Number\ of\ outcomes\ favorable\ to\ E}{Number\ of\ all\ possible\ outcomes}$$

Probability of an event that can occur in any one of n mutually exclusive and equally likely ways

Suppose that a *favorable outcome* is an **even** number. Then in one roll of a die, there are **three** favorable outcomes each of which is mutually exclusive. Therefore, P(even number) = 3/6 = .5

If you roll a die six times, an even number might not appear on exactly three occasions. It is also probable that if you roll a die 101 times and count the number of times an even number appeared, it would not be 50% of the time. There is a difference found in an experiment and the calculated probability.

Example

Two fair coins are tossed. Determine the probability that
a) two heads will occur.
b) at least one head will occur.

The sample space for the experiment is: **S = {HH, HT, TH, TT}**
Since the four outcomes are mutually exclusive and equally likely, the probability of each event is 1/4.
a) **P(HH) = .25**
b) **P(HH, HT, TH) = 3/4 = .75**

Based on these calculations, if you repeated this experiment 60 times, how many times would you expect to get two heads? Discuss.

Group Work

The experiment is drawing one card from a regular deck of 52 cards.

a. What is the probability of drawing a **five?**
b. What is the probability of drawing a **face card?**
c. What is the probability of drawing a **two** and a **three?**
d. What is the probability of drawing a **diamond** or a **three?**

Counting

For experiments like *rolling a die* or *picking one card*, it is not difficult to determine the number of possible outcomes for any experiment. But that is not always the case. In fact, one of the more difficult tasks is often to determine the number of possible outcomes. In order to be able to consider experiments beyond the very simple ones, you will have to master some *counting techniques*. These counting techniques are simply a way of arranging the information in order to count the number of possible outcomes. Consider two experiments in the drawings on this page.

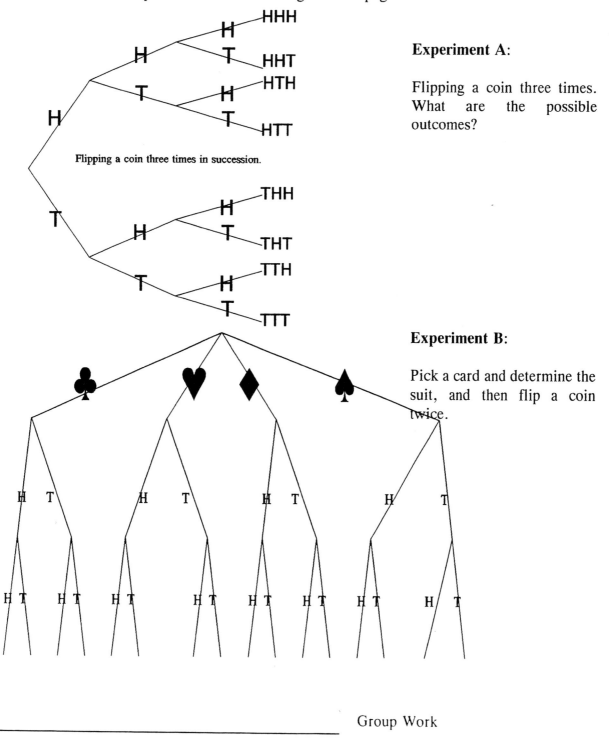

Flipping a coin three times in succession.

Experiment A:

Flipping a coin three times. What are the possible outcomes?

Experiment B:

Pick a card and determine the suit, and then flip a coin twice.

Group Work

List the outcomes of Experiment B.

Use the outcomes in Experiments A and B to determine the following.

Experiment A

1. How many different events are possible?

2. Is each event mutually exclusive?

3. What is the probability of getting exactly two heads?

4. What is the probability of getting at least two heads?

5. What is the probability of getting one head and two tails?

Experiment B

6. How many different events are possible? Is each event mutually exclusive?

7. What is the probability of getting a club, and two heads?

8. What is the probability of getting a heart and at least one tail?

9. What is the probability of getting at least one tail?

Types of Probability

Considering these experiments, two types of probability have been discussed: empirical and theoretical.

> Empirical Probability: when probabilities are obtained from experimental data. For example, an assembly line at Square D might make 1450 starters this month. If six are defective, then the probability of the line producing a defective starter is 6/1450 or .0041.

7.12

> Theoretical probability: when probabilities are obtained by logical reasoning and/or mathematical computation. For example, the probability of drawing a six of clubs is known since only 1 of 52 cards in a deck is a six of clubs: 1/52 = .0192.

7.13

Here is an example of empirical probability.

A manufacturing line produces switches that must turn off electricity on demand. About 1200 are produced every week. In an effort to improve quality, 75 are selected at random and tested. Only two are defective. What is the probability of getting a defective switch off the line? In this case there is no way to determine the theoretical probability for a defective switch. Empirical procedures must be used. Since only 2 of the 75 tested are defective, we conclude that 2/75 or 2.7% of the switches made on this line are defective.

Practice

1. A jar contains 97 marbles. 47 are blue, 30 are red, and 20 are green. Let the experiment be *picking a marble from the jar.*
   a. What is the probability of picking a red marble, P(red)?
   b. What is P(blue)?
   c. What is P(green)?
   d. What is P(green or blue)?
   e. What is P(purple)?

☺2. Let the experiment be picking a card from a regular deck of cards.
   a. What is P(eight)?
   b. What is P(an eight or less)?
   c. What is P(spade or seven)?
   d. What is P(three and spade)?

☺3. A spinner is marked so that the arrow will point to either A, B, or C. The areas are divided into fractions of 1/4, 1/3, or 1/2. Find P(A), P(B) and P(C), P(A or B), and P(A or C) for each spinner.

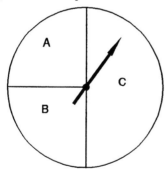

a.

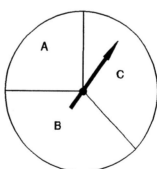

b.

4. A roulette wheel has 36 numbers (1 through 36) of which 18 are red and 18 are black. In addition to these 36 numbers, the numbers 0 and 00 also appear. These numbers, 0 and 00, are neither black nor red.
   a. Find the probability that the roulette wheel will come up with a red number. If you bet on red and it comes up red, the house pays you even money. Is that fair? Explain.
   b. Find the probability that the roulette wheel will come up with a 36. If you bet on 36 and it comes up 36, the house pays you $35 for every $1 you bet. Is that fair? Explain.

5.   A campus organization sold 1555 tickets for a raffle.
   a.   What is the probability that you, holding one ticket, will win the raffle?
   b.   A department bought 435 tickets. What is the probability that department will win the raffle?

6.   A supermarket did a survey of the number of people waiting in line at different times during the day from 5 until 8 in the evening and found the following probabilities:

Number in Line	0	1	2	3	4	5	6+
Probability	.08	.16	.26	.22	.15	.05	.08

   a. What is the probability there will be at most two people in line?

   b. What is the probability there will be at least four people in line?

☺7.   The histogram below shows the number of people arriving at a party during specific time intervals.

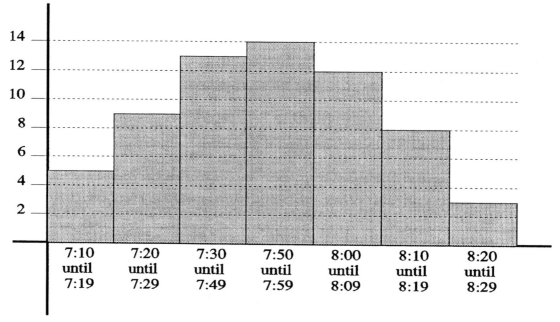

If one of the 64 people who attended the party is selected at random,

   a. what is the probability that person arrived at the party before 8:00?

   b. what is the probability that person arrived after 7:29?

   c. what is the probability that person arrived between 7:29 and 8:09?

8. Consider each experiment and determine if the events are mutually exclusive and if the probability is empirical or theoretical.

Experiment 1: The experiment is testing light bulbs to see if they burn or not. The sample space is made up of the events: {on, off}. After testing 200 bulbs, it was found that $P(on) = .98$ and $P(off) = .02$.

Experiment 2: The experiment is picking people from a group. The sample space is made up of the events: {male, female}. It is known that there are 40 people in the group and that 15 are female. Thus, $P(male) = .625$ and $P(female) = .375$.

Experiment 3: The experiment is polling people in a large city. The people are asked what electronic equipment they own in their home. The sample space is made up of the events: {TV, VCR, radio, CD player, computer, telephone answering machine}. Based on a sample of 100 homes, the probabilities were $P(TV)=.98$, $P(VCR)=.9$, $P(radio)=.92$, $P(CD\ player)=.6$, $P(computer)=.6$, $P(answering\ machine)=.55$.

9. Construct a picture to arrive at the different outcomes for the experiment *rolling two dice.*

How many different ways can the two dice end up; that is, the dice can end up with a 1 on the first die and a 1 on the second die (1 way), a 1 on the first die and a 2 on the second die (2 ways), a 1 on the first die and a 3 on the second die (3 ways), ..., a 3 on the first die and a 1 on the second die..., a 6 on the first die and a 6 on the second die? [Hint, the answer is more than 30.]

Complete the table below to show the number of different ways that each sum can appear as a result of the experiment. For example, a sum of five can be achieved four different ways: by a 1 on the first die and a 4 on the second, or 2 and 3, or 3 and 2, or 4 and 1.

Sum	2	3	4	5	6	7	8	9	10	11	12
Number of Ways				4							
Probability											

What is the sum of all the probabilities? What does this mean?

How do these answers explain the difference in the two histograms at the beginning of this section?

7.2 Area and Probability - Introducing the Normal Curve

Objectives of this section. The student will be able to
- *find the percent of area within a region of a histogram*
- *find the probability of an event from a histogram*
- *understand the influence that the standard deviation and mean have on the shape of the normal curve*

The health class at Poe High School surveyed all the students in the ninth grade to determine their height. The first histogram shows the heights of the males over the range of 48 to 80 inches. The second histogram shows the heights of the females over the range of 44 to 74 inches. Notice that the shape of the histogram in both cases is similar. This "bell" curve is of great significance and will be studied in this section.

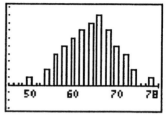

Males

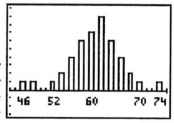

Females

Note: Histograms demonstrate the relative frequency of events. Therefore the height of a histogram does not represent the value of the variable. In the example above, people's heights are being measured but the height of the histogram represents the number of persons at a particular height. This distinction is important as we talk more about histograms.

Histograms

The *normal distribution* appears in many applications of physical science, sociology, biology, business, psychology and others. The normal distribution is frequently the result whenever there is a large number of independent trials or experiments. In order to understand the normal distribution, we will review functions in the context of a histogram. Consider a function whose dependent variable is the area in a histogram <u>above</u> the value of the independent variable.

It is better explained by an example. In the histogram to the right, the values of the independent variable range from 1 to 6. For each value of x there are a certain number of squares shaded in the column above it.

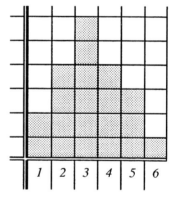

Example One

Area Under a Histogram

Define the function A, so that $A(x)$ represents the number of squares (**the area**) that is above that value of x. So $A(1)=2$ since there are two squares shaded above *1*. Then $A(2) = 4$ since there are 4 squares above 2. The other values of A are: $A(3) = 6$, $A(4) = 4$, $A(5) = 3$ and $A(6) = 1$. The picture to the right shows

$$A(x \leq 3) = A(1) + A(2) + A(3) = 2 + 4 + 6 = 12.$$

That is, the area above or to the left of x = 3 is 12.

A(x ≤ 3)

Example Two: Use the histogram to the right to determine the value of A(x).

a. A(3)

b. A(5)

c. A(8)

d. A(x ≤ 7)

e. A(3 ≤ x ≤5)

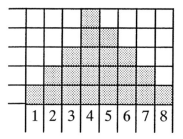

A(x)

Area Under a Curve

The *area under a curve* represents a different problem than finding the area within a histogram. First, there are no rectangles and, therefore, the area is not a polygon. Second, the values of *x* in A(x) are *discrete values*. However, a curve is *continuous*. In the group work above, you can find A(4), but can you find A(4.3)? No! The example above is *discrete*, not *continuous*.

The Normal Curve

Often, measured data in nature, such as heights of individuals in a population, are represented by a random variable whose area function may be approximated by the bell-shaped curve shown here. The curve extends indefinitely to the left and right and never touches the x-axis. The *normal curve* is the graph of one of the most important functions used in statistics.

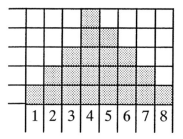

NORMAL CURVE

The graph (or shape) of the normal curve depends on two parameters: **the mean, μ; and the standard deviation, σ.** The mean determines the x-value of the maximum point. In terms of the example of heights, the mean height should be the height of the most people and is where the histogram is normally the highest. So too, the normal curve is highest at the mean. The standard deviation determines the width and the height of the normal curve. The greater the standard deviation, the wider and shorter the curve will be. In our example, if the standard deviation is small, then most people will have heights "close" to the mean and the histogram will not be very wide. If the standard deviation is large, then the histogram will be wide. The same can be said for the normal curve.

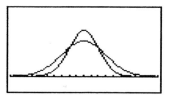

Same Mean

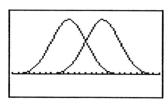

Same Standard Deviation

The figures illustrate two normal curves with the same mean and then two curves with the same standard deviation. The normal curve extends indefinitely in either direction. That means, in this context, that it is possible to have very short or very tall people, but it is very improbable.

The shape of the normal curve is dependent on σ and μ. The figure to the right illustrates how the standard deviation affects the curve. The relative position of the points $\mu + \sigma$, $\mu - \sigma$, $\mu + 2\sigma$, etc., always appear as they are shown. That means that μ is always in the middle of the curve. It also means that $\mu - 3\sigma$ will be at the left-most part of the curve with very little *area* under the curve and to the left of that point. It also means that most of the **area under the curve** is between $\mu - \sigma$ and $\mu + \sigma$.

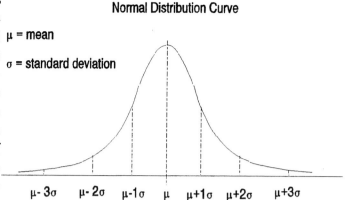

Normal Distribution Curve

μ = mean

σ = standard deviation

μ- 3σ μ- 2σ μ-1σ μ μ+1σ μ+2σ μ+3σ

The **area under the curve** is an important concept. In order to understand this concept with the normal curve, go back to **Example One (shown here)** at the beginning of this section. The function discussed in that example was A(x), the area above the value of x. Now consider another function, *P(x)*. P(x) will be associated with the **percentage of the total area above x.** The area of this histogram is 20 since there are 20 squares (2 + 4 + 6 + 4 + 3 + 1). Since there are two squares above x = 1, then P(1) = 2/20 = .1. Also, P(2) = 4/20 = .2; P(3) = 6/20 = .3; P(4) = 4/20 = .2; P(5) = 3/20 = .15; and P(6) = 1/20 = .05.

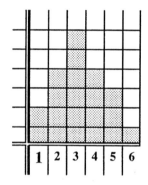

The percentage for x ≤ 3 would be the percentage of the area above or to the left of 3:   P(x ≤ 3) = (2+4+6)/20 = .6.   Similarly, P(x ≤ 5) = 19/20 = .95; P(3 ≤ x ≤ 5) = (6+4+3)/20 = 13/20 = .65.

This function, P(x), is also associated with probability. Suppose an urn contained 20 marbles: two with a 1 written on them, four with a 2 written on them, six with a 3 written on them, four with a 4 written on them, three with a 5 written on them, and one with a 6 written on it.

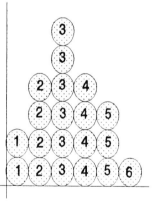

What is the probability that you would pick one marble and it have a 1 written on it? We know theoretically that would be 2/20 since there are two *successes* out of the available 20 *trials*. That fraction, of course, is the same fraction as P(1). The probability that you would pick a marble with a 2 written on it is 4/20 = P(2). The probability that you would pick a marble with a 3 written on it is 6/20 = P(3), and so on.

What about the probability of picking a marble with either a 1, 2 or 3 written on it? Since there are 12 such marbles, the probability is 12/20 or .6. That is: P(x ≤ 3) = .6.

Use Example Two in the group work to find the following.
  a. P(3)
  b. P(7)
  c. P(x ≤ 3)
  d. P(3 ≤ x ≤ 6)

This section has discussed area under curves and associated that area with probability. It also introduced you to the normal curve. In the next section, after practice with areas under curves, you will examine the normal curve further.

Practice

1. Consider the two histograms in the introduction to this section, males and females. For which data is the standard deviation likely to be greater?   For which set of data is the mean greater?

☺2. Before the individual results of a test were handed back to the students, the instructor posted a histogram showing the grade distribution for the entire class. That histogram appears to the right.

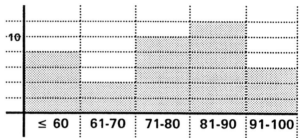

| | ≤ 60 | 61-70 | 71-80 | 81-90 | 91-100 |

  a. How many students received grades for the test?

  If any student, not knowing what their grade was, looked at the histogram of the grades
  b. what is the probability that their grade is between 61 and 70?
  c. what is the probability that their grade is between 71 and 100?
  d. what is the probability that their grade is 70 or less?
  e. Why do the answers to (c) and (d) add to 1?

☺3. Students in a PE class were told to run one mile. The time for each student was recorded and all times were to be made into a histogram. Use the data below to construct two histograms, one with wide class intervals and the second with smaller class intervals.

  Data: 250, 256, 259, 261, 263, 265, 266, 267, 270, 270, 271, 273, 275, 275, 276, 278, 278, 278, 279, 280, 281, 282, 283, 283, 284, 285, 286, 288, 289, 290, 291, 293, 296, 299, 301, 306, 311

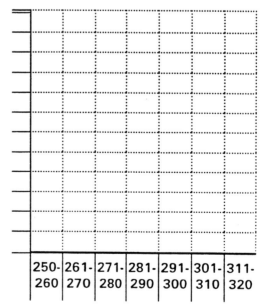

| 250-260 | 261-270 | 271-280 | 281-290 | 291-300 | 301-310 | 311-320 |

What overall affect does reducing the class interval have on the shape of the histogram?

Use the second histogram to answer the following questions.

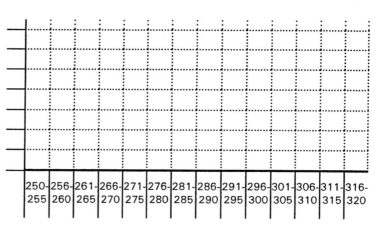

b. P(276-280)

c. P(x ≤ 280)

d. P(271 ≤ x ≤ 290)

e. P(x ≥ 291)

4. Matching. Match the information provided about the mean and standard deviation of a normally distributed population to the curve that best portrays that data.

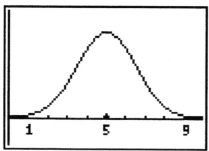

A.

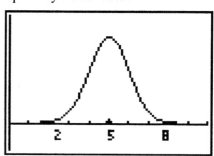

B.

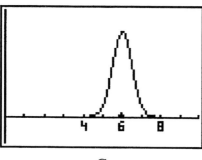

C.

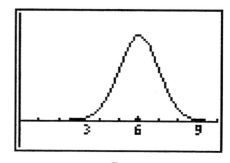

D.

i. $\sigma = .6, \mu = 6$

ii. $\sigma = .9 , \mu = 6$

iii. $\sigma = 1 , \mu = 5$

iv. $\sigma = 1.3, \mu = 5$

7.3 The Normal Curve, An Application of Probability

Objectives of this section. The student will be able to

■ *describe the affect on the probability of an event as its relative position is further from the mean*

■ *quote the approximate percentages for events with one, two, or three standards deviations from the mean*

■ *use a table of values to compute the probability of an event using the z-score*

■ *find the probability of events in the context of an application*

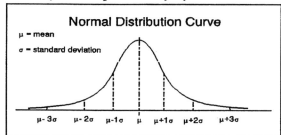

The normal curve is named after the German mathematician Carl Friedreich Gauss (1777 - 1855). Some consider Gauss to be one of the greatest mathematicians of all time. Many theorems and proofs are named after him.

The Normal Curve

The probability of an event occurring can be associated with a geometric area in a probability distribution or relative frequency distribution. The last section showed that the area under a curve and the probability of an event occurring were equivalent. The *normal curve* is a mathematical model of a probability distribution that occurs quite often. Many populations, especially those found in nature, are represented by a normal curve. The shape of the curve, bell-like, is determined by the standard deviation and mean of the population.

Remember:

everything isn't normal!

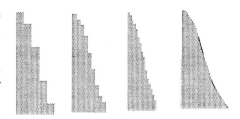

The normal curve is really the result of a constructing a histogram for a very large population and using very small class intervals. As the class intervals get smaller and smaller, the squared edges of the histogram gradually get rounded into a smooth curve. These figures demonstrate the transformation from frequency polygon to curve described above. The equation for this curve is shown to the right. But do not let this equation intimidate you. It will not be used.

As the size of the class interval decreases and the number in the sample increases the frequency *polygon* is slowly transformed into a smooth curve.

$$f(x) = \frac{1}{\sigma\sqrt{2\pi}}\, e^{(x-\mu)^2/(2\sigma^2)}$$

Facts About the Normal Curve

You need to know several facts about the normal curve (seen above). The area under the curve equals one. The curve extends to the right and left indefinitely; although, the curve is very close to the x-axis about three standard deviations from the mean. Notice that the curve is symmetric about the mean; $x = \mu$. That is, the height of a point

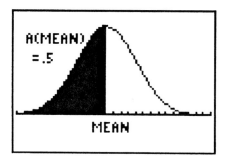

on the curve d units to the left of $x = \mu$ is the same as the height of a point on the curve that is d units to the right of $x = \mu$. Because of this symmetry and the fact that the area under the curve is one, the area to the right and left of the mean must be ½.

The values $\mu - \sigma$, $\mu + \sigma$ $\mu - 2\sigma$, etc., are important values. They not only identify points on the x-axis but also separate the area under the curve into specific regions. Approximately 68% of the area under a normal curve is within one standard deviation of the mean ($\mu \pm \sigma$). In other words, the probability that x will lie within one standard deviation of the mean is approximately .68.

$$P(\mu - \sigma < x < \mu + \sigma) = 0.68$$

$$P(\mu - 2\sigma < x < \mu + 2\sigma) = 0.95$$

$$P(\mu - 3\sigma < x < \mu + 3\sigma) = 0.997$$

The figure below illustrates these facts.

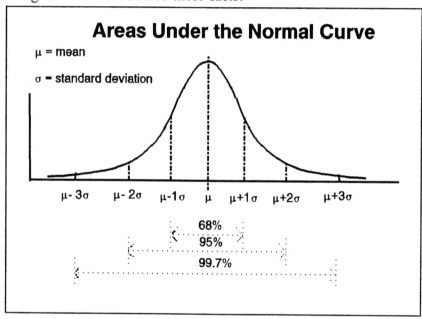

Area Under the Normal Curve

Example

Let x be a random variable whose values are the test scores obtained on a nationwide test given to high school juniors. Suppose, for modeling purposes, x is normally distributed with a mean of 800 and a standard deviation of 200. The probability that x lies within one standard deviation of the mean is 68%; that is, **P(600 ≤ x ≤ 1000) = .68.** This means that 68% of the scores fall between 600 and 1000. The probability that x lies within two standard deviations of the mean is 95% and **P(400 ≤ x ≤ 1200) = .95.** Therefore, 95% of the scores fall between 400 and 1200. Similarly, **P(200 ≤ x ≤ 1400) = .997** and 99.7% of all scores fall between 200 and 1400.

Area Table

To use the normal curve to determine probabilities, you must know the areas between <u>any</u> given limits. Since the equation of the curve is known, this has been calculated. The table to the right is the result of some of these calculations.

The values in the table are given "within z standard deviations of the mean." This means that each area given is that which lies between the mean and z *standard deviations on one side* of the mean (see the figure at the bottom of the page). If you think of the units for z as standard deviations from the mean, the formula shown here makes a lot of sense. To find out how many standard deviations a score is from the mean, subtract the mean from the score and divide by the standard deviation.

$$z = \frac{x_i - \mu}{\sigma}$$

The z-score for x_i given σ and μ for a random variable, x.

The number z is called the **standardized variable** or **normalized variable.** Remember that the curve is symmetrical about the mean so the values apply to either the left or right side.

The figure to the right goes along with the table. To find the area of the shaded region, simply look up the z-score in the table corresponding to the right most limit of the area. Remember, the figures in the table are true on either side of the mean. That means that
$$P(\mu - 1 \leq x \leq \mu) = P(\mu \leq x \leq \mu + 1)$$

For example, find A(2). A(2) represents the area between the mean (a z-score of zero) and a z-score of 2 (see the figure to the right). Using the table, we find that A(2) = .4772 or about 48%.

x	AREA	x	AREA
0.1	.0398	2.1	.4821
0.2	.0793	2.2	.4861
0.3	.1179	2.3	.4893
0.4	.1554	2.4	.4918
0.5	.1915	2.5	.4938
0.6	.2258	2.6	.4952
0.7	.2580	2.7	.4965
0.8	.2881	2.8	.4974
0.9	.3159	2.9	.4981
1.0	.3413	3.0	.4987
1.1	.3643	3.1	.4990
1.2	.3849	3.2	.4993
1.3	.4032	3.3	.4995
1.4	.4332	3.4	.4997
1.5	.4332	3.5	.4998
1.6	.4452	3.6	.4998
1.7	.4554	3.7	.4999
1.8	.4641	3.8	.4999
1.9	.4713	3.9	.5000
2.0	.4772	4.0	.5000

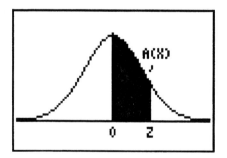

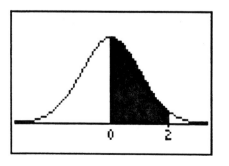

Find the z-score using the information given.

The population has a mean of 80 and a standard deviation of 10. Find the z-score for

(a) 75,

(b) 100.

The population has a mean of 8 and a standard deviation of .5. Find the z-score for

(c) 7.4,

(d) 9.2.

Find A(x) using the table.

(e) A(1.3)

(f) A(2.2)

Finding the Probability of an Event Within an Interval

The following scenario will set the stage for two additional examples using z-scores and the table.

A machine puts syrup into a bottle as the first step in producing the final mixture. Through empirical observation, it is found that the machine is currently placing an average of 2.1 oz. of syrup into the bottle. It is operating with a standard deviation of .05 oz.

If a bottle is selected at random, what is the probability it will contain
    a.   between 2.05 oz. and 2.175 oz.?
    b.   between 2.125 oz. and 2.175 oz.?

    (a)   The first step is to find the z-score for each value.

$$z(2.05) = \frac{2.05 - 2.1}{.05} = \frac{-.05}{.05} = -1 \qquad z(2.175) = \frac{2.175 - 2.1}{.05} = \frac{.075}{.05} = 1.5$$

The next step is to sketch a picture of A(x) under the normal curve. The total area is the sum of the areas: from the mean to z = -1, and from the mean to z = 1.5. Therefore:

$$z(-1) + z(1.5) = .3413 + .4332 = .7745$$

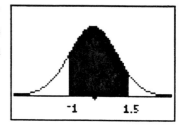

and the probability that a bottle, picked at random, will contain between 2.05 and 2.175 oz. is 77.45%.

(b)   The first step is to find the z-score for each value.

$$z(2.125) = \frac{2.125 - 2.1}{.05} = \frac{.025}{.05} = .5 \qquad z(2.175) = \frac{2.175 - 2.1}{.05} = \frac{.075}{.05} = 1.5$$

The next step is to sketch a picture of A(x) under the normal curve. The total area is the difference between the two areas: the area from the mean to z = 1.5 minus the area from the mean to z = .05. Therefore:

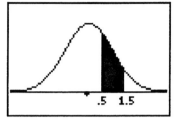

$$z(1.5) - z(.5) = .4332 - .1915 = .2417$$

and the probability that a bottle, picked at random, will contain between 2.125 and 2.175 oz. is 24.17%.

Application

Problem: The heights (in inches) of the 1500 male adults are normally distributed with $\mu = 70$ and $\sigma = 4$. How many males in the group are over six feet tall?

Solution: The first step is to find the z-score for 72 (six feet):
z (72) = (72 - 70) / 4 = .5.

The next step, as always, is to sketch the area you want to find. In this case we notice that the area we wish to find is *to the right of z =* **.5**. Remember: (1) the table provides areas from the mean left or right, and (2) the area to the right of the mean is .5. The area you wish to find is the area to the right of the mean, **less** the area between the mean and .5 (striped-area). Therefore P(x ≥ .5) = .5 - .1915 = .3085. This calculation means that about 31% of the population will be over six feet tall. To find the actual number of males taller than six feet, multiply: .3085 * 1500 = 463.

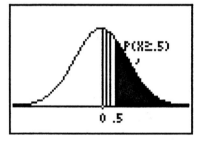

In conclusion, based on the information provided, 463 of the 1500 males in this population will be taller than six feet.

☺ 1. Find the z-score based on the information provided.
    a.   Find z(32) if $\mu = 25$ and $\sigma = 5$.
    b.   Find z(22) if $\mu = 26$ and $\sigma = 9$.
    c.   Find z(12) if $\mu = 25$ and $\sigma = 10$.

2. Find the area under the normal curve
    a.   between the mean and $z = 1.4$.
    b.   between the mean and $z = -2.2$.
    c.   between $z = .7$ and $z = 1.3$.
    d.   between $z = -.8$ and $z = .3$.
    e.   greater than $z = 2$.
    f.   less than $z = -3$.

3. The length of time a generator will run of four gallons of fuel is normally distributed with $\mu = 6$ hours and $\sigma = 15$ minutes. Find the probability that the generator will run more than 6 hours and 30 minutes.

☺ 4. A company sells 24 ounce bottles of cooking oil. A machine is used to place the cooking oil into containers. The machine is designed so that $\mu = 24.2$ oz. It is determined that machine has a standard deviation of .08 oz. What is the probability that the machine will place **less than** 24 ounces in any given container?

☺ 5. A company that tests water expects to see some bacteria, even in good water. The number of bacteria for a certain size sample of good water is expected to have $\mu = 30$ with $\sigma = 5$. What is the probability that any sample will have more than 50 bacteria? What should be concluded if a sample came in with more than 50 bacteria?

6. 2000 families were surveyed in a community. The family income was normally distributed with $\mu = \$50,000$ and $\sigma = \$10,000$.
    a. How many families have incomes from $40,000 to $60,000?
    b. How many families have incomes less than $24,000?
    c. How many families have incomes between $62,000 and $73,000?

7.4   The Exponential Function

Objectives of this section.   The student will be able to
- ○   *evaluate exponential functions*
- ○   *sketch the graph of an exponential function*
- ○   *solve applications which involve exponential functions*

Rules for
Exponents

Rules for Exponents

1. $a^m a^n = a^{m+n}$

2. $\dfrac{a^m}{a^n} = a^{m-n}$

3. $(a^m)^n = a^{m \cdot n}$

4. $(ab)^m = a^m \cdot b^m$

5. $\left(\dfrac{a}{b}\right)^m = \dfrac{a^m}{b^m}$

6. $a^0 = 1$

7. $a^1 = a$

8. $a^{-1} = \dfrac{1}{a}$

Exponential Functions

Exponential functions have an important role not only in mathematics but also in business, economics, and other areas of study.   They involve a constant raised to a variable power.   Linear, quadratic, and polynomial functions are examples of what are called algebraic functions.   An algebraic function is a function that can be expressed in terms of algebraic operations alone.   If a function is not algebraic, it is called a *transcendental function*.   Exponential functions are transcendental functions.

The function f defined by
$$f(x) = b^x$$
where b > 0, b $\neq$ 1, and the exponent x is any real number, is called an **exponential function** with base b.

Exponential
Function

The main thing to notice about exponential functions is that the variable is an exponent.   We shall begin our study of exponential functions by examining the function: . . . . $f(x) = 2^x$.
First, we complete a table of values for f(x).

$f(-2) = 2^{-2}$

$= \dfrac{1}{2^2}$

$= \dfrac{1}{4}$

To find f(-2), we substitute -2 for x and evaluate.   Recall that 2 to a negative power is equal to 1 over 2 to the positive power.

Similarly,   f(-1)   can   be   evaluated.
Completing the chart:

$f(0) = 2^0 = 1,$

$f(1) = 2^1 = 2,$

$f(2) = 2^2 = 4$, and

$f(3) = 2^3 = 8.$

$f(-1) = 2^{-1}$

$= \dfrac{1}{2^1}$

$= \dfrac{1}{2}$

x	y
-2	
-1	
0	
1	
2	
3	

The graph of **f(x) = 2ˣ** is shown here along with the graphing window used.

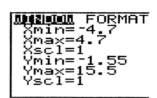

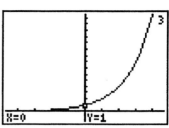

Group Work

Consider the exponential function **g(x) = 5ˣ**.

Complete the table
to the right and
sketch the graph
of this function.

x	y
-2	
-1	
0	
1	
2	
3	

Observations about bˣ

Consider several observations about the graphs of **b**ˣ where **x** is positive. These graphs have the same general shape; they rise from left to right--as x increases so does f(x). Each graph has a y-intercept of (0, 1) and extends indefinitely left and right. As the curves move to the left they get close to the x-axis since the values of f(x) get very small (**ex: f(-8) = 2⁻⁸ = .003906**). As the curves move to the right, they move steadily upward since the values of f(x) get very large (**ex: f(8) = 2⁸ = 256**). The shape of the graph of an exponential function where **b > 1** will <u>always</u> have these common characteristics.

It is common for people to confuse x^2 and 2^x. Don't; they are very different functions. The following set of figures illustrate the differences.

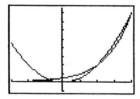

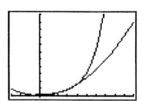

The first graph shows the two functions compared at values close to zero. But as the x-values get larger the differences between the two curves gets larger. The graphs illustrate this but the tables provide a better understanding of this difference. Notice that the curves intersect several times!

Applications

This text shall only consider exponential functions with b > 1 because our applications involve only that kind of functions. It should be said that many other applications involve exponential functions with **0 < b < 1**.

The first application involves compound interest. When a *principal* amount is invested and the interest earned is reinvested so that it too earns interest, the interest is compounded. There is interest on interest.

For example, consider investing $100 at 6% compounded annually. At the end of the first year the value of the investment is the beginning principal ($100) plus the

$100 + $100 (.06) = $106

$106 + $106 (.06) = $112.36

interest earned on the principal ($100*.06). At the end of the second year, the value of the investment is the principal at the end of the first year ($106) plus the interest earned ($106 * .06). These calculations are shown to the here. The $112.36 is called the *accumulated amount* and includes $12.36 in interest.

Generally, if a principal of P dollars is invested at a rate r compounded annually, the accumulated amount after 1 year is

$$P + Pr = P(1 + r).$$

The amount after the second year is

$$P(1+r) + [P(1+r)]r = P(1+r)(1+r) = P(1+r)^2.$$

This pattern continues. After three years the compound amount is $P(1+r)^3$. In general, the compound amount A, of a principal P at the end of t years at a rate r compounded <u>annually</u> is given by the formula:

$$A = P (1 + r)^t$$

Example: Suppose $100 is invested at 6% compounded annually. Find the accumulated amount at the end of seven years. How much interest was earned?

The compound amount can be found from the formula:

A = $100 (1 + .06)^7
A = $100 (1.06)^7
A = $100 (1.50363)
A = $150.36

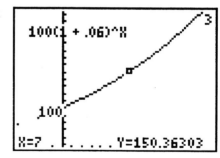

The interest earned is the difference between the accumulated amount and the initial principal: $150.36 - $100 = $50.36.

Compound Interest Formula

If the interest is earned in smaller increments of time another version of the formula is used.

The accumulated amount A of a principal P compounded at a periodic rate i at the end of n interest periods can be calculated with this formula. i is the quotient of r over m where r is the annual rate and m is the number of interest conversion periods in one year.

$$A = P(1 + i)^n$$

where $i = \dfrac{r}{m}$

and m is the number of periods in one year.

Example: Find the accumulated amount of $1500 at 8% compounded monthly at the end of 15 years. In order to use the formula we first find i and then n.

$i = \dfrac{.08}{12}$ The quoted annual rate, 8%, is divided by 12 the number of conversion periods in one year if compounded monthly.

$i = .006667$

$n = 12 \cdot 15 = 180$ n is the number of conversions over the 15 years: 15 years at 12 per year.

The accumulated amount is found to be $4960.38. That means that more than $3460 was earned in interest over the 15 years.

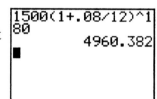

Group Work

1. Find the accumulated amount and the earned interest on an investment of $10,000 at 9.5% compounded annually over 8 years.

2. Find the accumulated amount and the earned interest on an investment of $5000 at 6% compounded quarterly over 12 years.

☺ 1. Complete the table of values for the function

$$f(x) = 3^x$$

and make a sketch of the graph of the function.

x	y
-2	
-1	
0	
1	
2	
3	
4	

☺ 2. Find the accumulated amount and the earned interest in each problem.

   a. $2500 is invested at 4% compounded annually over 5 years.

   b. $50,000 is invested at 5.5% compounded quarterly over 10 years.

   c. $124 is invested at 12% compounded daily over 2 years.

3. The value of $(1 + 1/m)^m$ as m gets large is an important constant. Use a calculator to complete the table.

$(1+1/12)^{12}$	$(1+1/365)^{365}$	$(1+1/1200)^{1200}$	$(1+1/1000)^{1000}$	$(1+1/10000)^{10000}$

Appendix

The following activities may be used to supplement the material in the text.

A Simulation

This activity can be done individually or with a group. Calculators may be used to perform the calculations, but spreadsheets would be preferable. They may be used to perform the calculations and draw the line graphs.

This activity is designed to strengthen the students' skills in using fractions and decimals and with organizing their work. It allows them to consider a simulation and how to change the simulation to meet other guidelines

Calculator Activities

This activity can be done individually or with a group. It is best to use groups if no one calculator is required in the class.

This activity is designed to strengthen the students' skills at using their calculator to perform calculations and to investigate functions numerically.

Ratios and Proportions

This activity can be done individually or with a group.

This activity is designed to strengthen the students' skills at investigating, organization and finding patterns. It will also strengthen their understanding of ratios and proportions.

A Numerical Study of Functions

This activity can be done individually or with a group. A spreadsheet is required for the activity and a graphing calculator is a plus.

This activity is designed to strengthen the students' skills at investigating functions and their skills at using spreadsheets.

A Simulation

Sometimes simulations are performed to model actual occurrences. Simulations are designed to demonstrate what could happen based on what is known about the situation.

The following simulation demonstrates a basic principal of statistics. Perform the simulation and try to identify and explain the basic principal underlying the project.

<u>Averaging Out</u>

You have always heard about the *Law of Averages*. But, do you know what that means (no pun intended)? This activity will help you develop your own thoughts about the *Law of Averages*.

This simulation can be used with several sports. I have chosen baseball.

Suppose that a batter's batting average for his career is .250. That means that he gets a hit 25% of his official at bats. So for his career, he may have 25 hits in 100 at bats or 101 hits in 404 at bats--his average would be .250.

Of course that doesn't mean that he will get 1 hit every four at bats. When players are hot they may average .450 for a while. They may also go "oh-for," which means they may not have a hit even after many at bats; like 0 for 12.

Let's simulate 30 games worth of statistics for a batter. We will need a die and a deck of regular playing cards. The die will be used to determine the number of official at bats the batter will get during one game. One to six at bats is not unrealistic. Once the number of at bats is determined. Draw the number of cards the die shows. A hit will be represented by a *club* type card. Since 25% of the cards are clubs, the batter's average should be .250. So (1) roll the die to determine the number of at bats and then (2) draw that number of cards with the number of clubs determining the number of hits.

We will examine three scenarios.
1. The batter is beginning a new year--he has no batting average (0 for 0).
2. The batter has started HOT! He has began the year batting .400, 16 for 40.
3. The batter has started COLD. He has began the year batting .150, 9 for 60.

Using different color ink to represent each scenario, record the cumulative average for the next 30 games for the player. In (2) and (3) begin with the given number of hits and at bats.

Place all three line graphs on **one** axis based on the <u>one set of data</u> you collect. Before you begin, discuss what you think will happen to the three curves and <u>write down your hypotheses</u>. After you have concluded the simulation, evaluate your hypotheses.

By drawing a diagonal line in the At bats and Hits cells, you can keep up with the cumulative totals easier. The last row in the table is computed by dividing the number of hits (cumulative) by the number of At bats (cumulative). The ordered pairs: (1,.333), (2,.200), (3,.300), (4,.308) ... will be graphed to make the line graph.

Sample data on the baseball average problem.

Games	1	2		3		4	
Hits	1	0	1	2	3	1	4
At Bats	3	2	5	5	10	3	13
Cumulative Average	.333	.200		.300		.308	

Scenario #1   Game Statistics

Games	1	2	3	4	5	6	7	8	9	10	11	12	13	14	15
Hits															
At bats															
Cumulative Average															

Games	16	17	18	19	20	21	22	23	24	25	26	27	28	29	30
Hits															
At bats															
Cumulative Average															

Scenario #2   Game Statistics

Games	1	2	3	4	5	6	7	8	9	10	11	12	13	14	15
Hits															
At bats															
Cumulative Average															

Games	16	17	18	19	20	21	22	23	24	25	26	27	28	29	30
Hits															
At bats															
Cumulative Average															

Scenario #3 — Game Statistics

Games	1	2	3	4	5	6	7	8	9	10	11	12	13	14	15
Hits															
At bats															
Cumulative Average															

Games	16	17	18	19	20	21	22	23	24	25	26	27	28	29	30
Hits															
At bats															
Cumulative Average															

Hypothesis:

Questions:

1. Did the batter bat .250 for his 30 games you played? If not, why not?

2. Suppose that this batter hits a home run in about 7.69% of official at bats. How could you find the number of home runs hit by this batter using this type of simulation?

3. Suppose another batter's average is .308. Can you devise a way to simulate his statistics for 30 games?

4. Suppose, based on your vast knowledge of the game, that there should be about twice as many games in which there are three or four official at bats as there are either 1, 2, 5, or 6 official at bats. How would you change your simulation?

Calculator Activities

Perform the following operations with a calculator and round as indicated.

1. Round to the nearest hundredth.

$$\frac{4.55^2 - \sqrt{27}}{3.2^3}$$

2. Round to the nearest thousands.

$$(3.5^2 + 5.3^{3.5})^3$$

Use estimation skills to estimate the answer prior to using a calculator to find the exact answer.

3. Estimate:

   Exact Value:

$$\frac{36.2 + 64.1}{\sqrt{16.1}}$$

4. Estimate:

   Exact Value:

$$4.2^2 - \frac{9}{5^5}$$

Find the exact value of the following expressions.

5. $$\frac{3\pi - 2\pi}{3}$$

6. $\sqrt{2} + \sqrt{3} + \sqrt{4}$

7. Find the largest value of the following expression: if the values of x, y and z can be 3,4,5 or 6 (each number can only be used once).

$$z(x^{-y})$$

8. Find the smallest value of the following expression: if the values of x, y and z can be 2,3,0 or -1 (each number can only be used once).

$$\frac{x^y}{z}$$

9. Find the smallest value of the following expression: if the values of x, y and z can be 2,3,4 or 5 (each number can only be used once).

$$\sqrt[x]{\frac{1}{y} + \frac{1}{z}}$$

10. Find the largest value of the following expression: if the value of x can be -2,4,-6 or 8.

$$\frac{x(1 + \frac{1}{x})}{\sqrt{x^2 + 2x + 1}}$$

Ratios and Proportions

Consider Figure 1. Measure the length and width of both the small and large rectangle. Calculate the ratio of the length to the width of the large rectangle and the ratio of the length to the width of the small rectangle. Compare the two ratios.

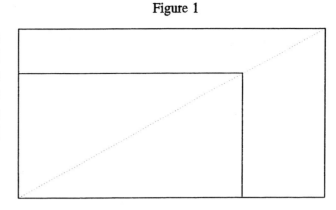

Figure 1

Consider Figure 2. Measure the length and width of both the small and large rectangle. Calculate the ratio of the length to the width of the large rectangle and the ratio of the length to the width of the small rectangle. Compare the two ratios.

Why do you think this comparison yielded these results? Explain.

If the rectangles were squares, what do you think the result would have been? Explain.

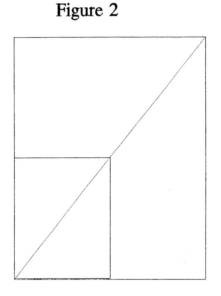

Figure 2

Consider Figure 3. Can you explain why comparison of the ratios of the length to the width are different than was found above?

Now consider the areas of the small and large rectangles in each figure. Can you find any measurement that is directly proportional to the ratio of the area of the large rectangle to the area of the small rectangle?

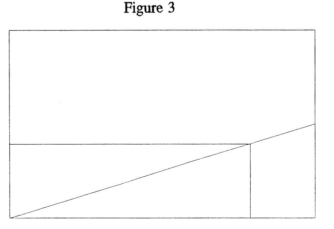

Figure 3

The Numerical Study of Functions

A spreadsheet is a useful tool in business/industry and can be very useful and meaningful if used properly in mathematics classes. The regional accrediting agencies now require students to demonstrate proficiency with the computer as graduates of a two-year college. Spreadsheets are a basic software tool that everyone should be capable of utilizing.

With regard to functions, using spreadsheets in the study of functions
- reinforces the meaning of variable,
- strengthens the concepts of independent and dependent variable,
- provides practice with order of operations,
- allows the student to visualize patterns without doing calculations, and
- provides an opportunity for some students to extend their investigations beyond the class' discussion.

Build a spreadsheet template to examine the values of a function.

^x	x	2x + 1	^2x + 1
1	1	3	2*A2+1
+A2+1	2	5	2*A3+1
+A3+1	3	7	2*A4+1
+A4+1	4	9	2*A5+1
+A5+1	5	11	2*A6+1
+A6+1	6	13	2*A7+1
+A7+1	7	15	2*A8+1

Examine these functions using a spreadsheet.

1.3*A2 - 3	1.3x - 3
(5*A2+2)/5	(5x + 2)/5
3-2*A2	3 - 2x

x^2 - 2x	+A2^2-2*A2
x^3 - x^2	+A2^3-A2
$\sqrt{(3-x)}$	@SQRT(3-A2)

The spreadsheet is useful for investigating small, or large, intervals of the domain. (Use this to "zero" in on a root, or to investigate the numerical values of a function around an asymptote, or to investigate the function on the outer limits of the domain.)

This is a spreadsheet that examines the values of $f(x) = \dfrac{1}{x-3}$ for values close to three.

		A		B		C
1	3	3	+A1	3	1/(B1-3)	error
2	.1	.1	+B1+$A$2	3.1	1/(B2-3)	10
3			+B2+$A$2	3.2	1/(B3-3)	5
4			+B3+$A$2	3.3	1/(B4-3)	3.333
5			+B4+$A$2	3.4	1/(B5-3)	2.5

This spreadsheet examines the extent of the function $f(x) = \dfrac{x}{x-3}$.

		A		B		C
1	20	20	+A1	20	B1/(B1-3)	1.1765
2	10	30	+B1+$A$2	30	B1/(B2-3)	1.1111
3			+B2+$A$2	40	B1/(B3-3)	1.0811
4			+B3+$A$2	50	B1/(B4-3)	1.0638
5			+B4+$A$2	60	B1/(B5-3)	1.0526

The values of the independent variable can be changed very simply in the two spreadsheets above by either changing the initial value (found in A1) or the increment (found in A2).

The initial value and increment for the independent variable can be changed easily. This allows the student to *investigate* changes in the dependent variable easily. Students can be given *templates* like these to investigate functions and/or formulas.

The TI-82 *table* feature allows a spreadsheet-like display. It cannot, though, be setup cell-by-cell like the regular spreadsheet. The display can achieve similar results.

The following pics display a function being selected, the table setup and the actual table from a TI-82.

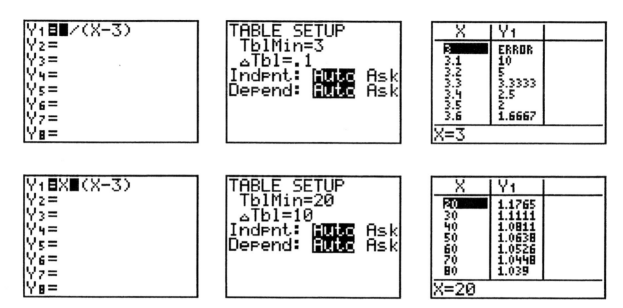

Spreadsheets are especially nice to use to investigate formulas with several variables. This next one can be used to find the accumulated value using compound interest at 6% monthly.....

		A		B		C
1	200	200	1	1	$A$1*(1+$A$2/$A$3)^B1	201.00
2	.06	.06	+B1+1	2	$A$1*(1+$A$2/$A$3)^B2	202.00
3	12	12	+B2+1	3	$A$1*(1+$A$2/$A$3)^B3	203.02
4			+B3+1	4	$A$1*(1+$A$2/$A$3)^B4	204.03
5			+B4+1	5	$A$1*(1+$A$2/$A$3)^B5	205.05
24			+B23+1	24	$A$1*(1+$A$2/$A$3)^B24	225.43

...then at 8% monthly.

		A		B		C
1	200	200	1	1	$A$1*(1+$A$2/$A$3)^B1	201.33
2	.08	.08	+B1+1	2	$A$1*(1+$A$2/$A$3)^B2	202.68
3	12	12	+B2+1	3	$A$1*(1+$A$2/$A$3)^B3	204.03
4			+B3+1	4	$A$1*(1+$A$2/$A$3)^B4	205.39
5			+B4+1	5	$A$1*(1+$A$2/$A$3)^B5	206.76
24			+B23+1	24	$A$1*(1+$A$2/$A$3)^B24	234.58

Build a spreadsheet to investigate the roots, extent and asymptotes of these functions.

+A2^2-9*A2-20	$x^2-9x-20$		$1/x$
	$12-x-x^2$	3/(A2-2)	$3/(x-2)$
	x^3-x		$x/(x^2-4)$

Take one of these functions, or another if you prefer, and a formula, and write an exercise for a student. Ask questions that cause the student to have to investigate and analyze the values of the function, make comparisons, and determine conclusions.

Index

angles 44
area within a histogram 180-2
area under a curve 181
arithmetic sequence 19
bias 52
box-and-whisker plot 118
circle 2
circumference 3
class 42
class interval 45
coefficient matrix 156
conformable 151
consistent 133
constraints 163
continuous functions 56, 73
coordinate axis 3
corner points of a region 162
counting 175
cube 35
decreasing 59
delta, Δ 32, 57
dependent 133
dependent variable 2, 6, 57, 71, 91, 127, 180
diagonals 34
dimension of a matrix 150
dimensional analysis 17
discrete functions 56
domain 70-72, 203
e 17,
elements 149
elimination 140
empirical probability 175
equality of matrices 150
equation of a line 83, 90, 94, 123
equilibrium 146-8
error 50
event 172
experiments 50-52, 172
exponential function 172, 191-3
exponents 10
feasible region 163
formula 1
frequency 42
fulcrum 106
function 1, 36, 69, 96, 102, 144
functional notation 3,7,123,148
general equation 94
geometric sequence 19

geometry 1
graph 3, 6, 73
Hero's formula 13
histograms 43, 45-48,66, 107, 178, 180, 185
identity matrix 152
inconsistent 141
increasing 57-61
independent variable 2, 6, 57, 71, 91, 127, 180
inequalities 8, 158-163
intercepts 89
inverse of a matrix 156
line of best fit 124
line graphs 3, 56
linear models 122
linear regression 128-9
linear programming 158, 163
linear equation 132, 136, 149, 162
logistic growth 18
margin of error 51
matrices 149
matrix multiplication 151
matrix equations 155
mean 104, 107, 119, 181
measures of central tendency 103
median 104
mode 104
model 123
model 3, 29, 97
mutually exclusive 172
negative exponent 18
normal data 113
normal curve 180
normal distribution 179
numeric zoom 145
numerical investigation 7, 144
objective function 163
optimal solution 167
optimum values 60
order of operations 14
ordered pair 2, 57, 132
origin 3
outlier 116
pattern 32
percent 5
pie charts 42
point-slope form 94
populations 36

probability 172
probability of an event 186
proportion 12, 30
Pythagorean theorem 13
quartiles 117
random sample 38
random numbers 39
range 70-72, 203
ratio 12
rearranging equations 79
rectangle 1
recursive formula 19
regression 122
relation 69
relative position 145
right triangle 13
right circular pyramid 2
rules of exponents 189
sample 36
sample space 172
sampling 36, 42
SCANS 54
scatter plot 3, 58
signed numbers 77
similar triangle 12
simulation 35, 196
slope 7, 80-6, 91
slope-intercept form 90, 94
solid 2
solution 132
sorting data 103
sphere 2
standard deviation 104, 181
subscripts 19
substitution method 138
sum Σ 110
system of linear inequalities 161
system of linear equations 132
table of values
template 127, 146, 202
theoretical probability 176
three-pronged approach 61
tracing 133
trend 3, 57
triangle 2
unit 11, 16
variability 50, 110
variables 2
voluntary selection 41
z-score 119, 187
zooming 133